Food and Us
The incredible story of how food shapes humanity

Food and Us
The incredible story of how food shapes humanity

By

Seamus Higgins
University of Nottingham, UK
Emails: seamus.higgins@nottingham.ac.uk; info@thebookfoodandus.com

Hardback ISBN: 978-1-83767-736-8
PDF ISBN: 978-1-83767-737-5
EPUB ISBN: 978-1-83767-738-2

A catalogue record for this book is available from the British Library

© Seamus Higgins 2025

All rights reserved

Apart from fair dealing for the purposes of research for non-commercial purposes or for private study, criticism or review, as permitted under the Copyright, Designs and Patents Act 1988 and the Copyright and Related Rights Regulations 2003, this publication may not be reproduced, stored or transmitted, in any form or by any means, without the prior permission in writing of the Royal Society of Chemistry or the copyright owner, or in the case of reproduction in accordance with the terms of licences issued by the Copyright Licensing Agency in the UK, or in accordance with the terms of the licences issued by the appropriate Reproduction Rights Organization outside the UK. Enquiries concerning reproduction outside the terms stated here should be sent to the Royal Society of Chemistry at the address printed on this page.

Whilst this material has been produced with all due care, The Royal Society of Chemistry cannot be held responsible or liable for its accuracy and completeness, nor for any consequences arising from any errors or the use of the information contained in this publication. The publication of advertisements does not constitute any endorsement by The Royal Society of Chemistry or Authors of any products advertised. The views and opinions advanced by contributors do not necessarily reflect those of The Royal Society of Chemistry which shall not be liable for any resulting loss or damage arising as a result of reliance upon this material.

The Royal Society of Chemistry is a charity, registered in England and Wales, Number 207890, and a company incorporated in England by Royal Charter (Registered No. RC000524), registered office: Burlington House, Piccadilly, London W1J 0BA, UK, Telephone: +44 (0)20 7437 8656.

For further information see our website at www.rsc.org

For general enquiries, please contact books@rsc.org

For EU product safety enquiries, please email books@rsc.org or contact Royal Society of Chemistry Worldwide (Germany) GmbH, Römischer Hof, Unter den Linden 10, 10117 Berlin.

Printed in the United Kingdom by CPI Group (UK) Ltd, Croydon, CR0 4YY, UK

Dedication

To my wonderful family: my amazing wife, editor-in-chief of concept and storyline; my son, Sé, thanks for the cover design; and my daughters, Jane, Julie Ann, and Jo, for their early proofreading.

For Mary, Ivy, Rosie, and Sé, our next generation of "A" young people, may this book help foster a positive relationship with our food systems and guide you in making the right decisions for personal health and sustainability.

I also dedicate this work to my friends, colleagues, and all those who tirelessly manage, work, and teach within this extraordinary industry. I invite you to reflect deeply on the myriad environmental, social, economic, cultural, and political factors increasingly shaping our endeavours. Likewise, recognise that our food system is not just a series of transactions; it is a vital network that impacts every single person living on this planet—a reality that has always been true and remains profoundly significant today.

Preface

As indicated on the cover sheet of this book, I have always had a passion for food and the food occasions that accompany it. But back in the early seventies, I never imagined that my entire career would revolve around the food industry. Similarly, when I joined the University of Nottingham as a course director to develop a new master's program in Food Process Engineering, I certainly did not foresee that I would eventually write a book on the same topic.

My career in food has also encompassed its production, origins, and some of the well-known foods and food brands we cherish today. Similarly, from a business perspective, how the significant growth of the food production industry has proven to be one of the most profitable financial investment sectors when compared to other industries.

At the start of the last century, Malthusian theory raised concerns that population growth might outstrip food supply, leading to famine, disease, and other crises. That never happened. Today, the global population has exceeded 8 billion, and our food industry continues to produce enough, and more, to feed nearly everyone.

However, is there still a grain of truth in Malthus's beliefs? Our recent advancements aimed at increasing food production and

---

Food and Us: The incredible story of how food shapes humanity
By Seamus Higgins
© Seamus Higgins 2025
Published by the Royal Society of Chemistry, www.rsc.org

profitability have also contributed to new environmental, sustainability, and health concerns. Moreover, 10% of the world's population still goes hungry.

The more recent growth of the food industry has also coincided with a significant shift in consumer eating habits. The way people around the globe now eat and drink has begun to conflict with our biology, resulting in notable changes in body composition.

Working from simple engineering principles, before one can consider a problem-solving approach or highlight some practical solutions that may account for constraints and trade-offs, an obvious question arises: how did all of this come about? What changed?

The exploration of that same question became the foundation for writing this book. Initially, my goal was to document the evolution of the food industry and its more well-known brands since the Industrial Revolution. However, as I delved deeper into the research, I realised that our food story actually began millions of years ago and is closely tied to our evolution from hominins to *Homo sapiens*. In fact, it was this same human evolution and our relationship with food that laid the groundwork for our modern society and the development of our current food system.

If we consider our evolution from hunter-gatherers, over millions of years, we have only been farming for about 0.003 per cent of that time. We have also only been consuming industrially produced food for approximately 0.00003 per cent of that same period. Hence, the mismatch we are now experiencing between the food we eat and our brains and bodies, which are still hardwired and designed for survival, just as they were in earlier times.

As the eventual working title for the book became "Food and Us, Still Evolving," I reflected that both of these elements were the two major contributors to answering the question of what had changed. Indeed, our incredible food history and subsequent evolution have continued to shape humanity since one of our earliest known ancestors, Lucy, began walking on two legs approximately 3.2 million years ago.

Our food future? As discussed in the final chapters of the book, it is evident that our current global food system is complex and plays a crucial role in both human health and the sustainability of our planet. However, it is also clear that our food system is no longer sustainable in its present form, and our global future will depend on addressing both of these essential aspects.

On a more positive note, we have the necessary resources to implement the changes required to create a more sustainable food system. Financially, the investment required would be less than the costs incurred during the recent COVID-19 pandemic. From a sociotechnical perspective, we also possess the skills, creativity, and expertise needed to develop and support a sustainable food environment that benefits both society and the health of our planet.

As Johan Rockström, a co-developer of the planetary boundaries framework, recently stated, "The global food system holds the future of humanity on Earth in its hands."

Seamus Higgins

Contents

Chapter 1	**Introduction**	1
	References	8
Chapter 2	**From Hominins to *Homo sapiens***	10
	2.1 "The Origin of the Species"	10
	2.2 From Early Hominins to *Homo sapiens*	14
	2.3 *Homo-sapiens*/Hunter-gatherers Evolving	19
	2.4 Life as Hunter-gatherers	21
	2.5 Evolving Human Instincts and Food	22
	References	26
Chapter 3	**The 1st Agricultural (Neolithic) Revolution**	29
	3.1 Learning to Farm or Developing Religion?	29
	3.2 The School Fees of Agricultural Development	33
	3.3 Developing New Food and Taste Profiles	37
	References	41
Chapter 4	**Ceres, the Goddess of Grain and Agriculture**	43
	4.1 Grain Milling	43
	References	54

Food and Us: The incredible story of how food shapes humanity
By Seamus Higgins
© Seamus Higgins 2025
Published by the Royal Society of Chemistry, www.rsc.org

Chapter 5	"Give Us, This Day, Our Daily Bread"	56
	References	63
Chapter 6	Regional Food Develops	65
	References	70
Chapter 7	A Cornucopia of New World Food	72
	7.1 The Columbian Exchange	72
	7.2 Developing a Sweet Tooth	74
	7.3 Preserves and Condiments	77
	References	78
Chapter 8	The 20th Century: Industrialised Agriculture	80
	8.1 Farming	80
	8.2 Crops	82
	8.3 Livestock	85
	References	91
Chapter 9	The 20th Century and the Business of Food Production	94
	9.1 Food Production as "One of the Modern Arts"!	94
	9.2 As Advertised: Taste, Price and Convenience (HFSS)	97
	9.3 From a Nation of Shopkeepers to "Every Little Helps"	106
	References	110
Chapter 10	20th-century Food Advice	113
	10.1 Regulating the Industry	113
	10.2 Controlling CHD, Saturated Fat or Sugar?	117
	10.3 "The Times They Are a Changing"	124
	References	126
Chapter 11	The 21st Century: Diet Food, Profitability, and Environmental Concerns	129
	11.1 Diet Food	129

Contents   xiii

11.2	Profitability and a Changing Market	131
11.3	Environmental and Sustainability Concerns	137
	References	140

Chapter 12   Food and Us: Our Unique DNA and Physical Makeup   142

12.1	Genotypes and Phenotypes	145
12.2	Our Microbiome	147
12.3	Our Food and Other Animals	150
	References	152

Chapter 13   Divining Food Energy and Nutritional Intake   154

13.1	Processed Foods and Additives	154
13.2	From Steam Engines to Bomb Calorimeters and Food Energy Labels	156
13.3	Why NOVA?	160
13.4	Dopamine: The Pathway to Pleasure	165
	References	168

Chapter 14   Rethinking Our Food System's Status Quo   171

14.1	Food Supply, Money and Profit	171
14.2	Our Animal-sourced Meat Dilemma	178
14.3	Our Nomenclature: *Homo sapiens*, and Obesity	184
	References	191

Chapter 15   Towards a More Sustainable Future Food System   196

15.1	Costing the Change Required	196
15.2	Breaking the "Junk" Food Cycle	202
15.3	Supporting a Food System for a New Generation: Generation A (Alpha)	209
15.4	Scaling Up More Sustainable Technology and Innovation	214
	References	223

Epilogue   226

Subject Index   231

CHAPTER 1

Introduction

It was a Persian scholar by the name of Heraclitus, who coined the phrase, "the only constant is change", some 2500 years ago. He viewed ever-present change as a fundamental essence of our universe. Think of how that same change constant has developed for you during your lifetime and/or allowed you to take some time out to begin reading this book!

By way of our evolution, human change is also an integral part of that same change constant.

It has taken more than 3 million years since we learned to walk upright on two legs and evolved from hominins to our current species, *Homo sapiens*. Of course, all that change and subsequent evolution has been fuelled by just one key ingredient; food.

When we think of food, there are several ways to describe it: the taste, the experience, the feeling or, as our more recently created food industry would say, a product that hits our so-called "bliss point".[1] Food is one of the basic necessities of life and an integral part of our cultural identity. It makes us feel good, satisfies our hunger, nutritional needs and sometimes even our emotional needs. Indisputably, food matters for every single individual on this planet; it always has!

From the time we are born, feeding a baby is one of the first acts of love a mother bestows on her child. Growing up as

---

Food and Us: The incredible story of how food shapes humanity
By Seamus Higgins
© Seamus Higgins 2025
Published by the Royal Society of Chemistry, www.rsc.org

children, our food and nutritional intake or lack thereof has a lifelong impact on how we live, for better or worse.

As religion evolved, several cultures and scriptures called food "a gift from god(s)." From a Christian perspective, Genesis 1; 29. In ancient India, the combination of yoghurt and honey was known as the "food of the gods," just as ancient Mesoamericans believed that the cocoa tree was sacred, which is why chocolate became known as a food of the gods.

With what we know today about how our bodies have evolved and our dependent relationship between the food we eat and good health, perhaps the one ancient message that still rings through today is from Greek philosophers who advised us on the difference between *sitos,* the staple, and *opson,* the relish, and warned against overly prizing the latter!

Beyond sustenance, it was the start of food trading that created the world's first currency some 3500 years ago. Post the egalitarian nature of hunter-gatherer societies, the ability to generate and control food distribution by way of grain surpluses became a path to power and influence in agricultural communities. Food and control of its supply laid the foundations for the critical elements of today's contemporary economies.[2]

From a cultural perspective, food has also been a centrepiece of our traditions, celebrations, and folkloric festivities. Take the word Carnival and its association with public events such as parades or street parties; it originates from a folk etymology derived from the Latin words *Carne vale*, "farewell to meat". Since the 12th century, it is still celebrated today by its original proponents, the Venetians of Italy, before fasting for what was then their 40-day catholic fasting period of Lent.

So which came first, the chicken or the egg? A popular puzzle or icebreaker topic related to food origins. Of course, in evolutionary terms, the chicken's distant dinosaur ancestors were laying eggs long before the first chickens evolved. But in today's society, a similar question about food may not be easy to answer. Was it better food availability that helped us become more populous, or was it human ingenuity with increasing medical and scientific proficiency that helped us multiply and become the most populous mammals on earth?

Global population growth changed from a linear trend to an exponential model in the 19th century. The world's population has jumped eightfold in the past two centuries, from an estimated one billion to eight billion people. This defied all Malthusian projections of the time that a growing population would outrun a limited food supply, leading to a global downward spiral.[3]

The same period also saw the dawn of a new food era brought about by the Industrial Revolution. Once again, a changing food supply, led by industrialisation, changed a predominantly agricultural and rural population to a new industrialised and urban lifestyle.

Unlike other industries that developed over the same period, a key difference is that food production has always been about supplying a basic human need rather than creating a profitable income stream.

Thus began the nascent food industry's conflict with business management theories rooted in economics, which also evolved in the 20th century. The rise of capitalism and the use of other service-linked sectors, such as advertising, accountancy firms, asset investment agencies, and management consultancy. Masters of business consultancy, such as Frederick Winslow Taylor, Michael Porter, Tom Peters, and Michael Hammer, would have a disproportionate effect on business management theories in general, including a growing food industry.

What is a food company's purpose and management's role? Peter Drucker, a prolific contributor to the Harvard Business Review, stated in 1954, "there is only one valid definition of business purpose: to create a customer" with a more inclusive view of "stakeholder capitalism," including stakeholders, employees, and consumers.[4] Alternatively, is it more as per the economist and Nobel laureate Milton Friedman, who published his famous essay in The New York Times Magazine in 1970 arguing that the sole social responsibility of businesses is to increase their profits? [5]

By the end of the 20th century, forces manifested by globalisation, trade liberalisation, and urbanisation changed the nature of our food systems further by increasing the diversity and affordability of food and consolidating corporate ownership.

Today's modern food industry can best be defined as a complex, global collective of diverse businesses that supply most of the food and drink consumed by the world's population. The $12 trillion industry represents more than 10 percent of global consumer spending and 40 percent of employment.[6]

From a profitability perspective, if one looks at returns by economic sector between 1963 and 2014, consumer staples provided a 13.3 percent compound annual growth rate (CAGR) return on investment, outperforming automotive, aeronautical and even computer industries.

A $1000 stock investment in the food industry in 1963 would now be worth $1 million plus.[7]

While the food industry has achieved many successes over the past 50 years in terms of scalability and providing an abundant food supply to many parts of the world, these successes have not come without consequence or controversy.

With an industry focus developing on a commodity-driven, brand-valued, cost-driven profit model, in some cases, it even defies all nutritional food logic!

I.e. how did a pharmacist's cure for his morphine addiction become the world's most popular soft drink, selling over 10 000 soft drinks every second of every day globally? Alternatively, how did a pious American Seventh-day Adventist and eugenicist's cure for "heinous sin, self-pollution, or masturbation" become the most popular breakfast cereal in Ireland?

Similarly, with convenience food, one of our time's most iconic food logos is the McDonald's double arches or big yellow M on a red background.

In the 1960s, McDonald's hired the design consultant and psychologist Louis Cheskin to create a new logo. Cheskin is considered a pioneer in understanding how colours can affect a food brand's efficacy, positing that red triggers stimulation, appetite, and hunger and attracts attention, while yellow triggers feelings of happiness and friendliness.

He was also a subscriber to Freud's theories on how sexuality drives human behaviours and believed those impulses to be a valuable tool in marketing. As such, he claimed that the new logo would have a Freudian pull for customers, insisting that the double arches would transmit a subliminal message for "Mother McDonald's breasts."[8]

Whether it was the colours or Freudian association with the logo, extensive advertising and marketing spend, or increasing portions of cheap fast food (average portion sizes have more than doubled since McDonald's started in the 50s!), the McDonald's chain is now the largest restaurant company in the world. Globally, the fast food market is estimated at just under $1 trillion per annum. McDonald's is a market leader with an approximately 21 percent share, supporting over 40 000 outlets in over 120 countries worldwide.[9]

The end of the 20th century saw the Industrial Revolution end with the realisation that nature does not have an infinite capacity to absorb the world's waste, be it greenhouse gas emissions or food waste. Likewise, the realisation that the world does not have an endless supply of natural resources such as fossil fuels or minerals for fertilisers.

The Agri-Food sector has now become the world's second-largest emitter of greenhouse gases, responsible for nearly one-third of all human-caused emissions.[10]

A third of all edible food produced goes to waste, even though 10 per cent of the global community goes hungry.[11]

Along with environmental and sustainability concerns, the more recent growth of the food industry also coincides with a dramatic shift in consumer eating habits; how the globe now eats and drinks has clashed with our biology to create significant changes in body composition.

Obesity levels in the UK and the US have more than trebled in the last 30 years. The World Obesity Federation predicts that one billion people globally, including 1 in 5 women and 1 in 7 men, will be living with obesity by 2030.[12]

The cause of this rapid rise in obesity, and subsequent rise in Type 2 diabetes and other diet-related illnesses, is complex and has been blamed on, amongst others, modern lifestyles. Of course, high-calorie, energy-dense food and clever food marketing play a significant role.

Similarly, the more recent revolution in food science, modern grocery retailing, and the use of lower-cost calories and ultra-processed fats and carbs that cause protein dilution in the food supply have also contributed to the same problem.

While these "ultra"-processed food products increase consumer demand by working on our innate preference for calorie-rich foods, they also disconnect the brake on our appetite systems by decreasing dietary fibre. It is perfect for getting us to

Figure 1.1 The evolution of Man.[14] Reproduced from ref. 14 with permission from Elsevier, Copyright 2019.

eat and buy more but devastating for our longer-term health and well-being (Figure 1.1).[13]

The Ellen MacArthur Foundation puts a number on it: For every dollar spent on food, society now pays two dollars in health, environmental, and economic costs.[15] The World Food and Agricultural Organisation (FAO) calculates the global price of hidden health and environmental costs of our present agri-food systems to be at least $10 trillion annually.[16]

Whether these concerns are viewed from an economic, social, political, or individual perspective, our present food model needs to change direction.

Before, the food industry's growth and profit drivers were solely based on taste, price, and convenience. The challenge now is how we, as individuals, civil society, industry, and indeed, governments, ensure that a future food industry can adapt to new emerging drivers, such as health and wellness, social impact and experience, or environmental and sustainability concerns.

The book explores our remarkable food journey over the past 3 million years and details the subsequent physical and cultural evolution that has transpired because of it.

It also explores the rich history of how our favourite foods evolved and developed over the past millennia and how the more recent growth of the food industry and its products now clash with our biology, creating a significant mismatch for our body's physical evolution.

Are we still evolving? Well, yes! Humans constantly evolve and will continue to do so as we continue to reproduce successfully. Think of evolution as heritable change over time, or as Heraclitus said, the only constant is change.

We are also witnessing how the current rate of change in cultural evolution is now changing in decades rather than thousands of years as per our physical evolution. If one reviews some of our recent physical evolutionary changes from a time perspective, the chance of Steve Earl or Ed Sheeran meeting their blue-eyed Galway girl 6000 years ago could never have happened! Originally, we all had brown eyes, and a genetic mutation in a single individual in Europe 6000 to 10 000 years ago led to the development of blue eyes.[17]

From a food perspective, 4000 years ago, virtually no adult human could properly digest animal milk. However, through evolutionary change, northern Europeans began to inherit a genetic mutation that enabled them to do so.[18]

More recently, researchers have found another genetic variant in populations in India and East Asia that have favoured vegetarian diets over many generations.

This new genetic variant allows these people to efficiently process omega-3 and omega-6 fatty acids from a plant-based diet.[19]

Conversely, when one looks at more recent cultural evolution, that difference becomes personified with our shift from rural to urban, agriculture to commerce, isolation to interconnectedness, less to more technology and education, lesser to greater wealth and smaller families/household units. On a positive note, unlike the physical attributes of evolutionary change, a significant advantage of social change is the ability to change it and adapt to new circumstances.

As my earlier question asks about increasing medical and scientific proficiency helping us to multiply, could that same resource of sociotechnical innovation and immense human ingenuity assist us in achieving the change in direction our current food system requires?

Before outlining a more positive and sustainable possible future food scenario, the book also details a better understanding of our DNA makeup and how we measure the energy value and composition of the foods we eat. It also explores the intricate interplay between our genetic and physical makeup and our enteric nervous system and brain cells, which work in unison to regulate how our body digests food.

These intriguing new discoveries are revolutionising our understanding of how digestion, mood, health, and even the way we think are intricately linked to the food we eat.

Chinese is one of the oldest written languages in the world, with at least six thousand years of history. Unlike other cultures that use alphabetic systems, Chinese writing is logographic, meaning each symbol or character represents a word.

I have always found it remarkable that all those years ago, the originators of the Chinese character for vitality or "Jing" (精) chose the two characters "米" (rice), which would have been brown rice at that time, and "青" (plant or vegetables) to represent the same word.

It was also a Chinese philosopher, Confucius (500 BCE), who said, "Study the Past if You Would Define the Future."

I hope you enjoy your deep dive into "Food & Us" and trust that our future food choices can evolve further to create a new global vitality or "Jing" (精) for both ourselves and the only planet we have!

REFERENCES

1. P. Rao, R. L. Rodriguez and S. P. Shoemaker, Addressing the sugar, salt, and fat issue the science of food way, *npj Sci. Food*, 2018, 2, 12.
2. Chapurukha (Chap) Kusimba, When – and why did people first start using money? The Conversation, June 20 2017.
3. T. R. Malthus, *Principles of Political Economy Considered with a View to Their Practical Application*, 1820.
4. W. Kiechel, *The Management Century*, Harvard Business Review, 2012.
5. M. Friedman, A Friedman doctrine The Social Responsibility of Business Is to Increase Its Profits, New York Times, 1970.
6. D. I. Hefft and S. Higgins, Re-engineering the Food Industry: Where Do We Go from Here? *Applied Degree Education and the Future of Learning, Lecture Notes in Educational Technology*, Springer, Singapore, 2022.
7. J. Charalambakis, *A Whole New World for Brands*, Blacksummit Financial Group, 2020.

8. E. Schlosser, *Fast Food Nation: the Dark Side of the All-American Meal*, Mariner Books/Houghton Mifflin Harcourt, Boston, 2001.
9. Statista Statistic report on Mc Donald's, Article ID: did-14355-1, 2024.
10. A. H. Kassam and L. Kassam, *Rethinking food and agriculture: new ways forward*, Woodhead Publishing, Duxford, 2020.
11. FAO report, The State of Food Security and Nutrition in the World, 2022.
12. World Obesity Federation report, The World Obesity Atlas, 2022.
13. D. Raubenheimer and S. J. Simpson, *Eat like the animals: what nature teaches us about the science of healthy eating*, Houghton Mifflin Harcourt, Boston, 2020.
14. L. R. Juneja, A. Wilczynska, R. B. Singh, T. Takahashi, D. Pella, S. Chibisov, M. Abramova, K. Hristova, J. Fedacko, D. Pella and D. W. Wilson, Chapter 5 - Evolutionary Diet and Evolution of man, *The Role of Functional Food Security in Global Health*, Academic Press, 2019, pp. 71–85.
15. Ellen MacArthur Foundation report, Food and the Circular Economy – deep dive, 2019.
16. C. Ruggeri Laderchi, H. Lotze-Campen, F. DeClerck, B. L. Bodirsky, Q. Collignon, M. Crawford, S. Dietz, L. Fesenfeld, C. Hunecke, D. Leip, S. Lord, S. Lowder, S. Nagenborg, T. Pilditch, A. Popp, I. Wedl, F. Branca, S. Fan, J. Fanzo, J. Ghosh, B. Harriss-White, N. Ishii, R. Kyte, W. Mathai, S. Chomba, S. Nordhagen, R. Nugent, J. Swinnen, M. Torero, D. Laborde Debouquet, P. Karfakis, J. Voegele, G. Sethi, P. Winters, O. Edenhofer, R. Kanbur and V. Songwe, The Food System Economics Commission, Economics of the Food System Transformation Global Policy Report, 2024.
17. Emily, All people with blue eyes have one common ancestor, Open Access Government, 2022, Available from: https://www.openaccessgovernment.org/blue-eyes-pigment-common-relative/134687/.
18. B. Schindler, *Eat Like a Human*, Little, Brown Spark, 2021.
19. A. Bolotkova, Vegetarian populations carry metabolic mutation, BioNews 845, 2016.

CHAPTER 2

From Hominins to *Homo sapiens*

2.1 "THE ORIGIN OF THE SPECIES"

Exploring the origins of where and how we evolved goes back to the golden age of Greek philosophy, when Anaximander of Miletus theorised about the origins of our species in the sixth century BCE. Anaximander may have been the first natural scientist to describe principles about life's origins and the universe's creation based on natural forces rather than supernatural explanations.

In 1750 CE, Carl Linnaeus, a Swedish botanist exploring the same subject, devised a scientific system to classify all living things. He ascertained that humans and apes were similar enough to be classified together in the zoological order of primates.

Forty years later, James Burnett, an eccentric judge and philosopher from Scotland, suggested that humans were related to orangutans and that Africa was humanity's ancestral home.[1] Burnett is now credited with being one of the first scholars to introduce the concept of evolution.[1] It was in 1859 when Charles Darwin, the English naturalist, geologist and biologist, fired the ongoing debate when he published his now-famous book "On the Origin of Species."[2]

Contrary to common belief, Darwin did not include humans in his initial discussion of species evolution. He believed the

---

Food and Us: The incredible story of how food shapes humanity
By Seamus Higgins
© Seamus Higgins 2025
Published by the Royal Society of Chemistry, www.rsc.org

subject of human evolution was so "surrounded with prejudices" that he was determined to avoid it entirely! It was the philosopher Herbert Spencer[3] who began to formulate evolutionary ideas before Darwin's work appeared. He believed that the fundamental physical laws of evolution meant that progress of all kinds depended on struggle and competition. Spencer coined the phrase "the survival of the fittest" to describe evolution, a phrase Darwin only added to his writings at a later stage.

Frustrated with the debate on human evolution at the time, Darwin eventually published his two-volume book entitled "The Descent of Man and Selection in relation to Sex" in 1871.[4] This book was the first of Darwin's published works to contain the word 'evolution'. The book's first part applies his theory of evolution to the human species, and the second section explores the role of sexual selection in evolution. For Darwin, sexual selection also explained how different human races had developed.

By the time Darwin's theories gained traction, many Victorians recognised a vision of the world that seemed to fit their social experience in evolutionary thinking. The scale of change during the 19th century and the impact of industrialisation, urbanisation, and technological innovation on people's lives was unprecedented.[5]

The idea of a 'struggle for existence' central to Darwin's theory of biological evolution was a powerful way to describe Britain's competitive capitalist economy in which some people became enormously wealthy, and others struggled amidst the direst poverty. Evolution seemed to confirm this view: species compete and struggle, and only some – the fittest and best – survive. Darwin was also convinced that cooperation is essential, especially for creatures that live in groups, including humans. At the time, other scholars believed competition was key to development. Theories of cultural evolution had yet to make their mark!

Darwin's theory of evolution, change over time, was based on three central tenets that he defined as natural selection.

- Because resources are limited in nature and most species produce more offspring through reproduction than their environment allows, there is a constant struggle for existence.
- Individuals within species are different and show different variations. Those with advantageous features are not only

more likely to survive but are also more likely to pass on their attributes by way of sexual intercourse.
- This, in turn, leads to the principle of inheritance. As offspring are more likely to resemble their parents, these advantageous characteristics are passed on to future generations. Conversely, weaker attributes in a particular species do not survive.

The essence of Darwin's theory is that if all three of the above conditions struggle for existence, variation, and inheritance are present, then natural selection occurs. Today, most scholars agree that five forces of evolution theory have influenced human evolution: natural selection, random genetic drift, mutation, population mating structure, and culture.[6] The latter force having a much more significant influence as we adapted to agriculture and today's modern world.[7]

More recently, several developments have enhanced our understanding of early human evolution.[8] Raymond Dart's findings in 1924 were significant when the "Taung Child" was discovered in South Africa and estimated to be 2.8 million years old. Dart named this fossil Australopithecus africanus. Although it was considered controversial at the time, this discovery was the first evidence researchers had of early human bipedalism, meaning upright, two-legged walking. Furthermore, the discovery of the first human ancestor fossil in Africa provided concrete evidence that Africa is the cradle of humankind.

The finding of "Lucy" at Hadar in Ethiopia in 1974 verified the same theory. Archaeologists Donald Johanson and Tom Gray unearthed the fossilised bones of one of humanity's earliest known ancestors, who walked upright. She was named "Lucy" after the Beatle's song "Lucy in the Sky with Diamonds", which was seemingly played repeatedly during the excavation dig!

On an evolutionary timeline, Lucy lived about halfway between apes and humans, sharing characteristics with both. It is estimated that Lucy most likely died about 3.17 million years ago. Lucy's discovery marked another turning point in our understanding of human evolution.[9]

While the timeline of the evolution of upright walking is well understood, why hominids took their first bipedal steps is unclear. Charles Darwin explained in his book "The Descent of

Man" that hominids needed to walk on two legs to free up their hands. He wrote that the hands and arms could hardly have become perfect enough to have manufactured weapons or to have hurled stones and spears with an accurate aim as long as they were habitually used for locomotion.

The problem with this idea is that the earliest stone tools or spears don't appear in archaeological records until much later. Another theory considers the efficiency of upright walking. In the 1980s, Peter Rodman and Henry McHenry at the University of California suggested that hominids evolved to walk upright in response to climate change. As forests shrank, hominid ancestors descended from the trees to walk across grassland stretches separating forest patches. Using two feet was the most energetically efficient way to walk on the ground.

Humans walking on two legs consume only a quarter of the energy that chimpanzees use while "knuckle-walking" on all fours. According to Rodman *et al.*'s theory, the energy saved by walking upright gave our ancient ancestors an evolutionary advantage over other apes by reducing the cost of foraging for food.

More recent studies of dental micro-wear combined with stable isotope analysis of ancient tissue in preserved hominin teeth show that several species, including Lucy's, were expanding their diet at the time.[10] Instead of mostly eating fruit from trees, they included grasses, sedges, tubers, fruits, seeds, and nuts. This change in diet allowed them to range more widely and to travel more efficiently in a changing environment. Lucy herself may have been collecting eggs from a lake. Fossilised crocodile and turtle eggs were found near her skeleton, leading to suggestions that she died while foraging for them. As Yuval Harari points out in his book "Sapiens"; "exploring the evolution of *Homo sapiens* from an insignificant animal in a corner of Africa to being the master of the entire planet and the terror of the eco-system.[11]

Learning to walk upright also came with its own trade-offs. The skeleton of our primate ancestors had developed for millions of years to support a creature that walked on all fours and had a relatively small head. Adjusting to an upright position was quite a challenge, especially when the body's scaffolding had to support an extra-large cranium. In learning to walk upright, the vertebral column also changed, and the ensuing upright posture enabled the forelimbs to be freed from locomotion. Bipedalism resulted in skeletal changes to the legs, knee and ankle joints, spinal vertebrae, and toes.[12]

Humankind paid a steep price for its lofty vision and newfound bipedalism, facing issues such as backaches and stiff necks. Women experienced these challenges even more acutely. An upright posture required narrower hips, which constricted the birth canal, while at the same time, babies' heads were becoming larger. This increase in head size made childbirth significantly more dangerous for women. Those who gave birth earlier, when infants' brains and heads were still relatively small and flexible, had better survival rates and were able to have more children. As a result, natural selection favoured earlier births in humans.

Indeed, compared to other animals, humans are born prematurely when many of their vital systems are still underdeveloped. A horse or a giraffe can trot shortly after birth. Human babies are helpless and dependent for many years on their elders for sustenance, protection, and education.

Another hypothesis for human bipedalism is that it evolved due to differentially successful survival from carrying food to share with group members. That same theory also significantly contributes to humankind's extraordinary social abilities and unique social attributes, which have further developed through its hunter-gatherer families and tribal lifestyle.

Scientists have also linked the evolution of the human hand – unique to humans due to its lengthy opposable thumbs and nimble fingers – to the emergence of stone tools approximately 2.6 million years ago. A recent study combining ancient evidence from fossil fingers and thumbs with cutting-edge computer muscle modelling concludes that South African hominins possessed flexible, capable thumbs, much like ours, as far back as two million years ago.[13]

By the time Lucy was found, anthropologists had accepted that this newly named species, Australopithecus afarensis, was indeed an early human, not just an ape, and Lucy is one of the most famous examples of this species.

2.2 FROM EARLY HOMININS TO *HOMO SAPIENS*

For almost two-thirds of our early evolution, the size of our ancestors' brains fell within the range of those of other apes living today. Since then, the human brain has tripled in size, most of that growth occurring in the past two million years.[14]

Determining changes in brain size over ancient times is challenging because we do not have actual ancient brains to weigh. However, palaeontologists can measure the interiors of ancient skulls and examine the rare fossils that have preserved natural casts of these skull interiors.[15] For instance, the species of Lucy's fossil, Australopithecus afarensis, had skulls with internal volumes ranging from 400 to 550 millilitres. This is similar to the skull volumes of modern chimpanzees, which are around 400 millilitres, and gorillas, which range from 500 to 700 millilitres.

Homo habilis, the first of our species *Homo* who appeared 1.9 million years ago, saw a moderate increase in brain size, including an expansion of a language-connected part of the frontal lobe called Broca's area. The first fossil skulls of *Homo erectus*, 1.8 million years ago, had brains averaging a bit larger than 600 millilitres.[16]

From here, our species embarked on a slow upward passage, reaching more than 1000 ml by 500 000 years ago. Early *Homo sapiens* had brains within the range of those of people today, averaging 1200 ml or more. As our cultural and linguistic complexity evolved, dietary needs and technological prowess made significant leaps forward at this stage, and our brains adapted to accommodate these evolutionary changes.

Reinforcing the theory that you are what you eat – or perhaps highlighting an evolutionary irony – is that human brain sizes have shrunk over the past 10 000 years. One possible explanation for this trend is limited nutrition in populations that rely on agriculture. Conversely, industrial societies have seen an increase in brain size over the past 100 years, as childhood nutrition has improved and disease rates have declined.

Proponents of the so-called "Paleo" diet, named after the Palaeolithic or Stone Age period that lasted from roughly 2.8 million to 10 000 years ago, argue that eating more like our early human ancestors is better suited to our evolved bodies. Followers of this diet believe that modern food production represents a genetic mismatch for humans. Therefore, they encourage individuals to eliminate certain contemporary food items from their diets, such as dairy products, processed foods, and agricultural products. Essentially, they adopt a pseudo-stone age or caveman diet.

The "Paleo" diet offers a simplified view of the eating habits of our prehistoric ancestors, but this perspective is overly one-dimensional. Its strong emphasis on meat, mainly from modern feedlots, does not accurately represent the diverse array of foods or the types of game and carrion that our ancestors consumed. Additionally, this approach neglects the active lifestyles that helped humans and animals avoid heart disease and diabetes during that time.

Approximately 2.8 million years ago, the East African vegetation transitioned from a wetland-closed canopy-dominated habitat to a drier, open savannah grassland. This change in the natural environment was accompanied by a worldwide faunal change that included the spread of large grazing animals.[17] Thus, the foods available to our hominin ancestors in an open grassland environment differed significantly from those of the jungle/forest habitats that had been home for millions of years.

The increase in grassland savannahs that spread across Africa, along with the growing numbers of grazing herbivores, became an obvious new food source for hominins. It's no coincidence that the dawn of stone tools dates to around the same period, some 2.6 million years ago. Known as Oldowan, these sharp stone flakes were created by striking a hard stone against quartz, obsidian, flint, or any other rock whose flakes could hold an edge. They were first discovered at Olduvai Gorge in Tanzania. Oldowan artefacts have since been recovered from several localities in eastern, central, and southern Africa.[18]

Invariably, most of these sites also had several animal bone fragments with cut marks matching the same tools. Indeed, more recent archaeological finds in Kenya detail similar finds dating back to 3.2 million years BCE. Contrary to earlier hypothesis by way of Dart *et al.*, it is doubtful that early man such as *Homo habilis* and or other genus, at the time, suddenly became adept at big game hunting on the African plains. That would take another million or two years to evolve!

Homo habilis, "the handyman" as he was first named by the duo Mary and Lois Leakey, who discovered the Oldowan fossils in 1962, was no hunter. More likely attracted by circling vultures, it probably scavenged the leftovers from big kills, such as an antelope in a tree by a leopard or a large animal such as a wildebeest that lions had slaughtered. The sharp flaked stones of the time were used to slice meat from bones, and the larger hammer stones, also found at the same sites, could be used to bash open bones to access the fat-rich marrow and brains inside this carrion.

The earliest evidence of what we might call persistent carnivory – part of the intensification and expansion of meat-eating – comes from a site in Kenya, where more than 3700 animal fossils and 2900 stone tools have been discovered in three separate layers, dating back approximately two million years.[19] This archaeological and fossil evidence includes dozens of bones bearing cut marks and percussion marks. The indications are clear that early humans, most likely *Homo habilis* or *Homo erectus*, processed several animal carcasses during repeated visits at the same location.

Although meat, marrow, and fat, or what we can refer to as animal-sourced food (ASF), were initially modest sources of nutrition for hominins,[20] they gained in importance as access improved through aggressive scavenging and later hunting. From a historical and evolutionary perspective, it is clear that *Homo sapiens* indeed emerged with the anatomical and physiological equipment of a habitual rather than a facultative meat eater.

Over the next 1–1.5 million years, it was not just the human brain size that changed; the shift from fibrous plants to ASF, from both a terrestrial and aquatic origin, together with the use of tools, paralleled a decrease in teeth size and jawbones, a reduction in chewing muscles, and weaker maximum bite force capabilities. *Homo erectus* developed smaller molars and also began to spend a lot less time on feeding than would be predicted from body mass and phylogeny with apes (5 percent instead of an expected 48 percent of daily activity in *Homo sapiens*).[21]

As a protection against meat-borne pathogens, the human stomach has also evolved to be one of the most acidic in the animal kingdom, similar to other carnivores and scavengers. The same adaptation led to a small intestine comprising 56 percent of total gut volume and a shrinkage of the large intestine (and therefore fermentative capacity) to a mere 20 percent, which is the inverse situation of what is found in apes today.[22]

According to the 'expensive tissue hypothesis', the increase in brain size was made possible – under selective pressure for more cognitive capacity – by an overall reduction in the size of the energy-consuming gut as well as by the supply of energy and nutrients *via* ASFs (*e.g.*, iron, zinc, vitamin B12, choline, docosahexaenoic acid, fat, cholesterol).[23] Nicotinamide (vitamin B3) has also been cited as a key brain-trophic element in ASFs.

The brain's exceptionally high energy needs may also explain why humans – infants in particular – have higher body fat than non-human primates. While the brain of an adult primate consumes less than 10 per cent of the total resting metabolic rate, that same measure amounts to 20–25 percent in the case of anatomically modern humans.[24] As the supply of meat increased, so did the size of human brains, eventually leading to the evolution of *Homo sapiens* (Figure 2.1).

The earliest evidence for hunting technology, in the form of hafted spear points, dates back to about 500 000 years ago; complex projectile weapons only appeared much later at 71 000 BCE.[26] Persistence hunting has been suggested as a mode of hunting that would have been possible without advanced technology. But how

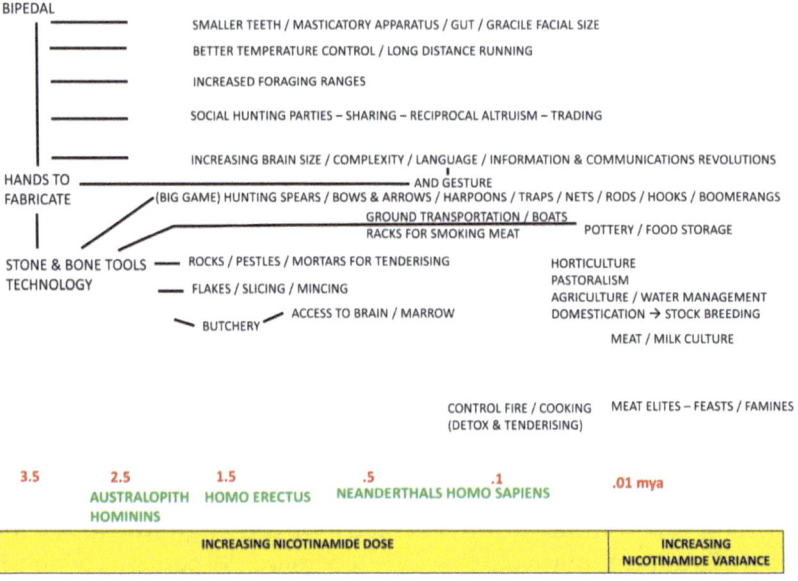

Figure 2.1 Our early evolution in a nutshell. Steadily increasing nicotinamide powered biological and cultural evolution.[25] Reproduced from ref. 25 with permission from Sage, Copyright 2017.

would one recognise this behaviour in fossil or archaeological records?

Although there is evidence for earlier traces of controlled fire use in East and Southern Africa, the earliest firm evidence of fire hearths in the form of burned seeds, wood, and flint, likely related to cooking, dates to about 800 000 years ago.[27] Early humans' manipulation of fire was another significant turning point for our ancestors; once fire had been domesticated, it provided protection from predators, offered warmth and light, and enabled the exploitation of a new range of foods.

Cooking helped tenderise meat and made plants more edible. Likewise, as brains evolved, hominids developed a more intricate knowledge of edible plant life growth and seasonal cycles. Examination of the Gesner Benot Ya'aqov site in Israel, which housed a thriving community almost 800 000 years ago, revealed the remains of 55 different food plants and evidence of fish consumption.[28]

With the introduction of spears, hunter-gatherers could now track larger prey to feed their groups. Modern humans were also cooking shellfish by 160 000 years ago, and by 90 000 years ago, they had developed specialised fishing tools that enabled them to haul in larger amounts of aquatic life.

2.3   *HOMO-SAPIENS*/HUNTER-GATHERERS EVOLVING

So, after three million years, various stages of evolutionary natural selection, random genetic drift, and mutation, people anatomically similar to us finally evolved around 300 000 years ago. While there is little agreement among experts on the question of whether to refer to all *Homo* species as human, there were approximately seven to eight, or more, different types of humans on the planet at that time.

Some of our now-extinct relatives, such as Neanderthals, are well known. The Neanderthals, *Homo neanderthalensis*, were stocky hunters adapted to Europe's cold steppes. The more primitive *Homo erectus* lived in Indonesia, and *Homo rhodesiensis* in central Africa. The related Denisovans, who were just unveiled this decade and discovered by ancient DNA, lived in Asia.

Another recent discovery: a large fossil skull found in China may belong to one of our mysterious, long-lost relatives, the

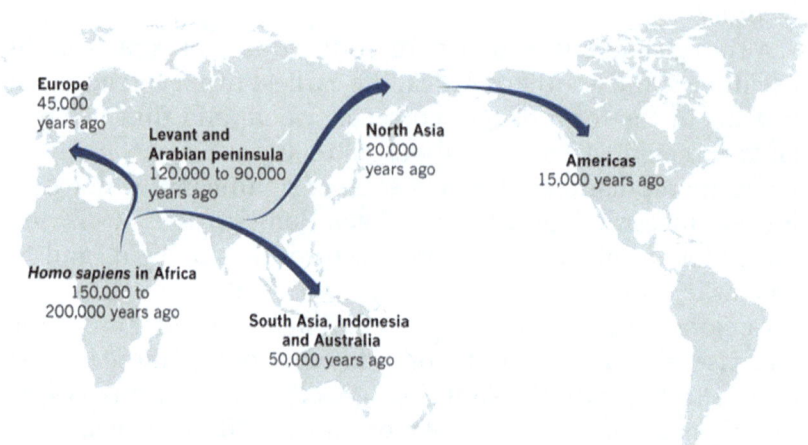

Figure 2.2 Human migration out of Africa.[31] Reproduced from ref. 31 with permission from Springer Nature, Copyright 2016.

Denisovans, potentially offering us our first glimpse of a Denisovan face. It has, however, been placed in a new human species – *Homo longi* – a name that derives from a Chinese term meaning "dragon" related to the adjacent river where it was found in Hebi province.[29]

More recent research, including new fossils, improved DNA research and dating techniques – confirm the complexity of modern human (*Homo sapiens*) origins.[30] Evidence now suggests that all modern humans are descended from an African population of *Homo sapiens* that spread out of Africa about 60 000 years ago but also shows that they interbred quite extensively with local archaic populations (Neanderthal and Denisovan genes are found in all living non-African populations, Figure 2.2).

Similarly, all descendants – Bantu, Berber, Aztec, Aboriginal, Tamil, San, Han, Maori, Inuit, and Irish – share certain peculiar behaviours absent in apes. All human cultures form long-term pair bonds between men and women to care for children. We sing and dance. We make art. We preen our hair and adorn our bodies with ornaments, tattoos, and makeup.[32]

While the precise details of how or when our tools, food, fashions, families, morals, and mythologies developed and varied from tribe to tribe and culture to culture, all living humans show the same behaviours. This suggests that these behaviours – or at least the capacity for them – are innate. They are the human condition, what it means to be human and the result of a shared ancestry that has evolved over three million years since we learned how to walk upright and lived together in hunter-gatherer tribes.

2.4 LIFE AS HUNTER-GATHERERS

Drawing on insights from various global sources, including archaeological, genetic, and cross-cultural research, as well as studies of contemporary hunter-gatherer communities, we gain some understanding of what human life was like during these periods. Our ancestors lived in small groups, similar to all primate species. Their social network could comprise up to 150 individuals, including family, friends, and acquaintances. This group was the tribe. The same tribe would spread out over a more extensive habitat by way of camps. Each camp consisted of some extended families, men, women and a few children, augmented by grandparents, cousins, uncles and aunts. The extended family members would move from camp to camp, sometimes as many as eight times a year, in search of areas with plentiful food, watering holes, and suitable camping spots.

Our ancestors lived with close and distant relatives from birth to death. With limited resources, these groups were egalitarian in nature. They foraged, ate, and slept together as nomads. Groups offered protection and allowed people to collaborate when they foraged, picked, gathered, and hunted. Sharing food was an essential survival strategy.

Thus, developing social relationships became extremely important. Indeed, unlike other primates, humans have developed eyes with a distinct colour contrast between the white of their eyes, known as the sclera, the coloured iris, and their black pupils.[33] The advantage of this is that one could immediately see what others in the group were looking at. In contrast, other primate species concentrate on how their heads are turned. Visual cues and gaze detection were crucial to human group behaviour, whether silently hunting together or discussing where and when to move camp.

To this day, human babies follow their parents' eye movements after approximately three months. If their mother looks to the right, they will do likewise.[34] This is the first speechless form of communication, as language only evolved recently, around 50 000–150 000 years ago.

Daily life consisted mainly of hunting, gathering and sharing food. Division of labour by gender became more pronounced with the advancement of hunting techniques, particularly for large game. Generally speaking, the men in the group kept themselves busy hunting, while the women collected herbs, grasses, berries, fruits, and nuts. As the evolutionary psychologist Ronald Giphart states in his book Mismatch, the term is generally used loosely here because, obviously, there were women who participated in hunting, just as there were men who stayed behind. However, it does suggest that gender roles are ancient. It may also explain why the brains of men and women, despite many parallels, are made up and function differently in some areas.

Evolutionary anthropologists assume that people at the time had, on average, more children than in present-day society, approximately four or five children per adult female. But infant mortality was high, and in the end, couples/family groups raised, on average, just over two healthy children into adulthood, thus creating a relatively stable population for many generations as these two children outlived their parents. Mothers breastfed their children for four to six years, and during this period, they were unlikely to conceive. Moreover, whilst trekking from camp to camp, parents could only carry two children, so for most of this period, population growth was held in check.

2.5 EVOLVING HUMAN INSTINCTS AND FOOD

Evolution, or change over time, occurs through the processes of natural and sexual selection. In response to environmental conditions, we have adapted both physically and psychologically to ensure our survival and reproduction. Hunting and gathering were humanity's original and most enduring competitive adaptations in the natural world, occupying nearly 95 per cent of our human history. Our thoughts, emotions, and physical characteristics have also been influenced by natural and sexual selection since prehistoric times.

Sexual selection theory describes how evolution has shaped us to provide a mating advantage rather than just a survival advantage. It occurs through two distinct pathways: intersexual competition and intersexual selection. Gene selection theory, the modern explanation behind evolutionary biology, occurs through the desire for gene replication. Evolutionary psychology connects evolutionary principles with contemporary psychology. It focuses primarily on psychological adaptations: changes in how we think to improve survival.

We evolved with psychological mechanisms, as evolutionary psychology now refers to them, to help us survive, and we still see traces of this in how our modern brain functions.[34] The past few decades and new developments in animal research, brain studies, genetics, and cognitive science have fundamentally changed our understanding of the brain, its inherent "software," and what triggers our response to specific environmental conditions.

We are primarily concerned with what we can perceive with our senses (sight, smell, taste, hearing, or touch). We are relatively less quickly alarmed by danger that is not immediately perceptible to our senses.[35] (e.g., climate change and its consequences). In divisive issues, our brain tends to let our self-interest prevail over that of others or the population at large. The interests of family members who are genetically related to us weigh more heavily than those of genetic strangers.

Our evolutionary brain also tends to be rather short-sighted. The future did not matter in the past because you had to survive on a day-to-day basis. Food could not be stored, so if you had a piece of meat, you would eat it, and leftovers would be shared with your family. If you saw a beehive in a tree or some ripe fruit, you would eat a little of the same before another human or animal could climb the tree. With the rise of lifestyle diseases, this short-term thinking can still be observed in modern society. People prefer to reward themselves in the short term (a tasty treat) rather than the long term (a healthy body).

Our palates, stomachs and even our DNA have evolved over millennia to like various food types, tastes and dislike others. In addition to dictating what food we enjoy eating, these psychological mechanisms also influence which partners attract us sexually, the type of leader we want to follow, and whom we can trust. The exact mechanisms can also be referred to as instincts as they influence our perception, thoughts, feelings, and actions, often at a subconscious level.

It has been more than 60 years since psychologist Walter Mischel sat a group of five-year-old children down at a table and gave them a simple choice: They could eat one marshmallow now, or if they waited a short time later, they could have two. Watching a video (plenty on U-tube, "The Marshmallow test") is always endearing as the children's dilemmas and reactions unfold to the same choice. One thing becomes apparent: the more exposure the children let themselves have to the marshmallow, the more likely they are to eat it. Kids who turned away from the marshmallow or distracted themselves in some way were more successful at waiting and receiving the second one.

Mischel and other psychologists argue that this behaviour is caused by the battle between two different brain systems: instant gratification (one marshmallow now) and long-term prudence (two marshmallows later). There is the Limbic system, the lower, more primitive part of the brain, which responds immediately and emotionally and allowed us as a species to survive a predator-filled environment in ancient times. Other parts of the brain, concentrated in the prefrontal cortex, allow us to do things like control our attention and think about the future and delayed gratification.

The marshmallow test has since become famous not because of our inherent instinct for instant gratification and self-control, but because of the striking way it also indicates the subject's longer-term life skills. Twenty years later, further research found a direct correlation between the number of seconds a child delayed eating the marshmallow and their eventual outcomes in early adolescence, such as SATS scores and/or how well the same child would do socially and cognitively in their teens.

The research also found that adults who had waited longer on the test as children reported being better at dealing with stress and frustration as adults and were also found to have lower body mass indexes.[36]

If we examine our more "primitive" brain side, the limbic system also acts as a control centre for conscious and unconscious functions; it regulates much of the body's activity. This set of brain structures plays a crucial role in our emotions, particularly those that evolved earlier and were essential to early survival. The limbic system also controls our autonomic nervous system. It supports non-conscious functions such as thirst, hunger, heart rate and regulating the body's internal clock.[37]

It also links to feelings of motivation and reward, learning, memory, the fight-or-flight response, and the production of hormones that help regulate the autonomic nervous system. In a later chapter, we will examine more closely how the same system is now creating mismatches with our current food supply.

Think of your fight or flight response to a dangerous or perceived immediate threatening situation. The same threat triggers a primal physical reaction in the body, leaving you breathless, with a pounding heart and a racing mind. From deep within your brain, a chemical signal speeds the release of stress hormones through the bloodstream, priming your body to be alert and ready to escape danger. Concentration becomes more focused, reaction time faster, and strength and agility increase. When the stressful situation ends, hormonal signals switch off the stress response, and the body returns to normal.

But in our modern society, stress doesn't always let up. Many of us now harbour anxiety and worry about daily events and relationships. Long-term activation of the stress system can have a hazardous, even lethal, effect on the body, increasing the risk of obesity, heart disease, depression, and a variety of other metabolic illnesses.

Hormones are the body's chemical messengers, sending signals from one area to another in response to environmental inputs and other information. The hypothalamus, part of the limbic brain, releases hormones that control many emotions, including pain, hunger, thirst, pleasure, sexual feelings, anger, and aggression.[38] It also helps the body maintain a state of homeostasis by regulating the autonomic nervous system. Homeostasis is the tendency of animals and humans to maintain relatively consistent internal environments by regulating temperature, metabolism, blood sugar, and other critical bodily functions.

More recently, the limbic system has been defined as a group of interconnected cortical and subcortical structures dedicated to linking visceral states and emotion to cognition and behaviour.[39] Again, as will be seen in later chapters, our food and nutrient intake is critical to our overall health and well-being. Hidden in the walls of our digestive system, this "brain in your gut" with more than 100 million neurons is revolutionising medicine's understanding of the links between digestion, mood, health and even the way you think. All of it depends on the food you eat.[40]

While our full evolutionary story is still a work in progress, all disciplines concur that, when it comes to natural selection and evolutionary change over the past three million years, it has been fueled by just one key ingredient: food. Food provides the energy needed for activity and growth. It also powers all bodily functions, such as breathing and maintaining a temperature that matches the environment's dictates. Consider our African savannah origins and the typical modern house temperature setting of approximately 20 degrees Celsius. It also relates to our body's mental and social health, as well as its ability to repair and maintain a healthy immune system.

Without air or water, which supplies the oxygen needed to give our cells the ability to break down food, or the latter, which allows us to digest our food and circulate the nutrients required by the bloodstream (blood is 90 percent water!), we could not survive. Water also plays an integral role by lessening the burden on our kidneys and the excretory system for the elimination of wastes produced by homeostasis.

Oxygen is a fundamental element, and water is formed by combining it with another fundamental element: hydrogen. When two parts of hydrogen combine with one part of oxygen, they create water (H_2O). Unlike our food supply, these base elements have remained unchanged for over 5 billion years. In contrast, the nature of food and our dietary habits have undergone significant evolution over time. Indeed, one could argue that food availability, or the lack thereof, has played a critical role in driving significant developments in human evolution.

In keeping with Heraclitus's theory of ever-present change, *Homo sapiens* were about to undergo another significant change as early agriculture evolved.

REFERENCES

1. S. D. Tucker, *Great British Eccentrics*, Amberley Publishing Limited, 2015.
2. C. Darwin, *On the Origin of Species*, Natural History Museum, S.L., 1859.

3. M. Francis, *Herbert Spencer and the Invention of Modern Life*, Routledge, 2014.
4. C. Darwin, *The Descent of Man, and Selection in Relation to Sex*, 1886.
5. D. Gange, *The Victorians*, Simon and Schuster, 2016.
6. National Academy of Sciences, *Systematics and the Origin of Species*, National Academies Press, 2005.
7. L. Mlodinow, *The upright thinkers: the human journey from living in trees to understanding the cosmos*, Vintage Books, New York, 2016.
8. J. Tomczyk, Facts and their interpretation in paleoanthropological enquiries, *Stud. Ecol. Bioeth.*, 2016, **14**(2), 115–132.
9. N. Wade, *The Science Times book of fossils and evolution*, Lyons Press, 1998.
10. F. E. Grine, M. Sponheimer, P. S. Ungar, J. Lee-Thorp and M. F. Teaford, Dental microwear and stable isotopes inform the paleoecology of extinct hominins, *Am. J. Phys. Anthropol.*, 2012, **148**(2), 285–317.
11. Y. N. Harari, *Sapiens: a Brief History of Humankind*, Harper Perennial, New York, 2011.
12. W. E. Harcourt-Smith and L. C. Aiello, Fossils, feet and the evolution of human bipedal locomotion, *J. Anat.*, 2004, **204**(5), 403–416.
13. T. L. Kivell, Human evolution: Thumbs up for efficiency, *Curr. Biol.*, 2021, **31**(6), R289–R291.
14. J. G. Fleagle, *Primate adaptation & evolution*, Academic Press, San Diego, 1988.
15. F. E. Grine, J. G. Fleagle and R. E. Leakey, *Springerlink (Online Service. The First Humans: Origin and Early Evolution of the Genus Homo*, Springer Netherlands, Dordrecht, 2009.
16. H. Jerison, *Evolution of The Brain and Intelligence*, Elsevier, 2012.
17. T. R. Mcclanahan and T. P. Young, *East African ecosystems and their conservation*, Oxford University Press, New York, 1996.
18. J. J. Shea, *Prehistoric Stone Tools of Eastern Africa*, Cambridge University Press, 2020.
19. M. Q. Sutton, *Archaeology*, Routledge, 2018.
20. P. S. Ungar, *Evolution of the human diet: the known, the unknown, and the unknowable*, Oxford University Press, Oxford, New York, 2007.

21. W. L. Jungers, *et al.*, The evolution of body size and shape in the human career, *Philos. Trans. R. Soc., B*, 2016, **371**(1698), 20150247.
22. M. Maxwell, *Human Evolution*, Columbia University Press, 1984.
23. H. Skorupa, *Outline and evaluation of the expensive-tissue hypothesis proposed by Aiello/Wheeler (1995)*, Grin Verlag Gmbh, München, 2008.
24. S. A. Heldstab, K. Isler, S. M. Graber, C. Schuppl and C. P. van Schaik, The economics of brain size evolution in vertebrates, *Curr. Biol.*, 2022, **32**(12), R697–R708.
25. A. C. Williams and L. J. Hill, Meat and Nicotinamide: A Causal Role in Human Evolution, History, and Demographics, *Int. J. Tryptophan Res.*, 2017, **10**, 1178646917704661.
26. R. B. Lee and I. DeVore, *Man the Hunter*, Routledge, 2017.
27. I. Tattersall, *Masters of the planet: seeking the origins of human singularity*, Palgrave Macmillan, New York, 2012.
28. D. Oates, *Of Pots and Plans*, 2002.
29. F. Romagnoli, F. Rivals and S. Benazzi, *Updating Neanderthals*, Academic Press, 2022.
30. J. Reader, *Missing links: in search of human origins*, Oxford University Press, Oxford, 2011.
31. R. J. Berry, *2 - No Primeval Eden*, Cambridge University Press, 2018.
32. N. Longrich, When did we become fully human? The Conversation, 2020.
33. J. G. Fleagle, *Primate adaptation & evolution*, Academic Press, San Diego, 1988.
34. R. Giphart and M. van Vugt, *Mismatch*, Hachette UK, 2018.
35. S. Pinker, *How the mind works*, W.W. Norton, New York (N.Y.), London, 1997.
36. W. Mischel, *The Marshmallow Test*, Random House, 2014.
37. J. B. Furness and M. Costa, *The Enteric Nervous System*, 1987.
38. W. Haymaker, E. Anderson and W. J. H. Nauta, *The Hypothalamus*, 1969.
39. M. Catani, F. Dell'Acqua and M. Thiebaut de Schotten, A revised limbic system model for memory, emotion and behaviour, *Neurosci. Biobehav. Rev.*, 2013, **37**(8), 1724–1737.
40. S. Ackerman, *Discovering the brain*, National Academy Press, Washington, D.C., 1992. Available from: https://www.ncbi.nlm.nih.gov/books/NBK234151.

CHAPTER 3

The 1st Agricultural (Neolithic) Revolution

3.1 LEARNING TO FARM OR DEVELOPING RELIGION?

The world's first agricultural revolution began around 11 000 years ago when human groups started working the land now referred to as the Fertile Crescent. This arc of land spans modern-day southern Turkey, Iraq, Syria, Israel and Lebanon. These Neolithic farming communities domesticated cereals such as emmer wheat, einkorn, wheat and barley.[1] They also grew legumes such as lentils, chickpeas, peas and flax and domesticated animals such as sheep, goats, pigs and cattle. Evidence of fig trees planted some 11 300 years ago has also been discovered in the Jordan Valley, which runs from the Sea of Galilee to the Red Sea.

Further evidence of such early settlements has been discovered throughout the Middle East, such as the Natufian culture that thrived in the Levant region, now Israel, Jordan, Lebanon and Syria around the same time. The Natufians were hunter-gatherers who subsisted on a diverse range of wild species. Still, they lived in permanent villages and devoted much of their time to the intensive gathering and processing of wild cereals. They built stone houses and granaries, storing grain for times of need. They also invented new tools, such as stone scythes for harvesting and stone pestles and mortars for milling.[2]

Food and Us: The incredible story of how food shapes humanity
By Seamus Higgins
© Seamus Higgins 2025
Published by the Royal Society of Chemistry, www.rsc.org

Over the next few thousand years, farming started to develop around the globe. Finding a convincing explanation for the reasons behind this change in human history on a global scale has been the subject of speculation for decades. Most researchers agree that there is no identifiable factor or obvious combination of factors for this transition, but plenty of evidence exists.[3]

Until recently, many scientists believed farming was brought to Europe by word of mouth. It was conjectured that European hunter-gatherers living near ancient farmers in the Near East observed more settled ways and disseminated this information to Europe. However, large-scale DNA genetic analysis has revealed that the first domesticated crops in Europe had a significant genetic contribution from the Near East and modern-day Turkey. This points to a migration of Near East farmers into Europe around 8500 years ago.

In Mexico, squash was the first crop domesticated, with cultivation beginning around 10 000 years ago. The first maize-like plants, derived from Teosinte, their wild ancestor, appear to have been cultivated in Mexico at least 9000 years ago. The first corncob as we know it today dates to around 4200 years ago.[4]

In South America, the first domesticated crops were potatoes, corn, squash, sweet potatoes, beans, yucca root and peanuts. Early Chinese farmers planted millet, rice and soybeans around 6000 BCE. The world's oldest known rice paddy fields, discovered in eastern China in 2007, also reveal evidence of ancient cultivation techniques such as flood and fire control.[5]

Whatever the reasons for this initial spread of farming practices across the globe, the resultant shift from a nomadic existence to a sedentary lifestyle sowed the seeds (pun intended) for our modern way of life.

In his 1874 book, "The Descent of Man", Charles Darwin stated that possessing property, a stable residence, and organising many families under a leader are essential requirements for civilisation. However, he added that the question of how civilisation evolved from "savages" to a more advanced state remains unresolved.

It was an American anthropologist and social evolutionist, Lewis Henry Morgan, who posited in 1871 that ancient societies progressed through stages from "savagery" to "barbarism" and finally to "civilisation".

Using more appropriate terminology today, archaeologists employ conceptual frameworks to explain the evolution of human life over the same period, based on their findings of ancient tools and technologies. Their organising concepts are the Stone Age, Bronze Age, Iron Age, *etc.* As Yuval Noah Harari states in his book Sapiens, this aligns more comfortably with their rational and mathematical methods. Therefore, when examining ancient periods, the materialist school dominates.

It was not until 1936, with the publication of Man Makes Himself, that V. Gordon Childe coined the phrase "the Neolithic Revolution." His second book, What Happened in History (1942), introduced a theory and synthesis that applied social models to the same archaeological data. The Neolithic Revolution not only marked the beginning of farming practices and permanent settlements, but it also led to significant societal changes, including the development of specialised roles, social hierarchies, and the emergence of complex civilisations.

As Richard Dawkins wrote in his seminal 1976 book The Selfish Gene, "There are some examples of cultural evolution in birds and monkeys, but... it is our species that really shows what cultural evolution can do".

The findings at a Neolithic site in south-eastern Turkey add another dimension to the debate on early farming and settlement.[6] Discovered by archaeologist Klaus Schmidt, Göbekli Tepe features massive carved stones dating back approximately 11 000 years, which were crafted and arranged by prehistoric people who had yet to develop metal tools or pottery. These megaliths predate Stonehenge in the UK, also built by early agricultural communities, by 6000 years.

The standing stones, or pillars, are arranged in circles with circle stones approximately 60 feet (18 metres) in diameter. The tallest pillars tower 16 feet (almost 5 metres) high and, as Schmidt says, weigh between seven and ten tons. While some stones are blank, others are elaborately carved; foxes, lions, scorpions, and vultures abound, twisting and crawling on the pillar's broadsides.

Schmidt's team found none of the tell-tale signs of settlement at Gobekli Tepe; no cooking hearths, houses, or trash pits and or none of the clay fertility figurines that litter nearby sites of about the same age.

Because Schmidt found no evidence that people permanently resided on the summit of Gobekli Tepe itself, he believes this was a place of worship on an unprecedented scale-humanity's first "cathedral on a hill".

Close to the temple's hill site, Schmidt's colleague Joris Peters, an archaeozoologist, identified more than 10 000 bone fragments of gazelle and other wild game such as boar, sheep, and red deer. He also found bones of several wild bird species, including vultures, cranes, ducks, and geese. He concluded, "It was pretty clear that we were dealing with a hunter-gatherer site."

Constructing a temple the size and scope of Gobleki Tepe would have taken many years and required a massive workforce with a steady food supply to enable its completion. To Schmidt and others, these new findings suggest a novel theory of civilisation. Scholars have long believed that only after people lived in settled communities did they have the time, organisation and resources to construct temples and support complicated social structures. However, Schmidt argues that it was the other way around; the extensive, coordinated effort to build these monoliths and the food resources required laid the groundwork for the development of complex societies.

Regardless of how and why it happened, the shift from a total reliance on wild resources to the use of domesticated foods and animals led to several fundamental and far-reaching changes in human society. The Neolithic Revolution involved far more than adopting novel ways of farming and food-producing techniques.

Over the next few millennia, it transformed the small, mobile groups of hunter-gatherers that had hitherto dominated human prehistory into sedentary societies based in built-up villages and towns.[7]

By approximately 9000-8500 BCE, the Middle East was dotted with permanent villages, including Jericho, one of the world's earliest continuous settlements. By about 8000 BC, the inhabitants of Jericho had grown into an organised community of approximately 2000–3000 people, capable of building a massive stone wall around the settlement, strengthened by enormous stone fortification towers.[8]

Over time, some of these farming societies transformed into much larger, more complex social systems characterised by cities, political states, and the start of class inequalities. The new way of life quickly expanded beyond its original zones through conquest and trade. Rulers and dynasties rose and fell, and the stone tools of archaeology made way for written documents as the primary source of evidence for human history.

3.2 THE SCHOOL FEES OF AGRICULTURAL DEVELOPMENT

Despite a commonly held belief that hunter-gatherers had a poor quality of life, an analysis of human skeletal remains from that period reveals that the advent of early agriculture did not necessarily result in positive progress for *Homo sapiens*.

3.2.1 Physical Changes

From a physiological perspective, adapting to agriculture led to changes that worsened physical conditions and increased infectious disease incidence in humans.

The nutritional standards of the growing Neolithic populations were inferior to those of hunter-gatherers. Several studies conclude that transitioning to cereal-based diets reduced life expectancy and stature and increased infant mortality and infectious diseases. Disease spread more rapidly in settlements than in hunter-gatherer societies. This was due to inadequate sanitary practices, people living together in close proximity, and the domestication of animals.

The first livestock was domesticated from animals that hunter-gatherers had previously hunted for meat. Domestic pigs were bred from wild boars, while goats came from the Persian ibex.[9] Larger domesticated animals made the hard, physical labour of farming possible. Milk and meat products added variety to the human diet, but the animal-sourced products carried infectious diseases, such as smallpox, influenza, and measles, which spread from domesticated animals to humans.

Some of the first domesticated farm animals included sheep and cattle. Sheep were domesticated around 10 000 years ago, and cattle around 8000 years ago in the Fertile Crescent. Water buffalo and yak were domesticated shortly after in China, India, and Tibet.

Other draft animals, including oxen, donkeys and camels, appeared much later—around 4000 BCE—as humans developed trade routes for transporting goods.

Multiple nutritional deficiencies, including vitamin deficiencies such as iron deficiency, anaemia, and mineral disorders affecting bones (such as osteoporosis and rickets), as well as teeth.[10] The average height decreased from 5'10" (178 cm) for men and 5'6" (168 cm) for women to 5'5" (165 cm) and 5'1" (155 cm), respectively. It was not until the twentieth century that the average human height returned to pre-Neolithic levels.

Salmonella enterica genomes recovered from human skeletons as old as 6500 years provide the first ancient DNA evidence supporting the hypothesis that the cultural transition from foraging to farming facilitated the emergence of human-adapted pathogens that persist to this day.[11] Bio archaeologists have also suggested a link between caries (tooth decay) and the transition from foraging to farming. Acid-producing bacteria in the mouth consume fermentable carbohydrates, which abound in wheat, rice, and corn.

To understand why modern-day humans' teeth are so prone to decay, we need to understand one's natural oral environment.[12] A healthy mouth is teeming with life! It is populated by billions of microbes representing over 700 different species of bacteria. Like us, having also evolved over several million years, most of these bacteria are beneficial and help the body fight disease, help with digestion, and regulate various bodily functions.

They produce alkalis and antimicrobial proteins that inhibit the growth of harmful species. Diets rich in carbohydrates feed acid-producing bacteria, thus lowering oral pH levels. This, in turn, affects our saliva balance, which buffers teeth against acid attacks by bathing them in calcium and phosphate to re-mineralise the surface. As we will see in a later chapter, the additional effect of sucrose—or sugar—on teeth enamel was yet to come.

3.2.2 Sociology

On the sociology side, the origins of agriculture and ensuing events also provide an excellent example for understanding the processes of cultural evolution[13] and how society rapidly changed from a predominantly egalitarian society to its antonym with the advent of elitism, religion and farm workers.

Agricultural food production laid the foundations for sedentary living, increasing population size, accumulating possessions, and the inevitability of a more complex society. As early farmers learned the intricacies of planting and growing grain, the domestication of livestock, and their acceptance of a link between hard work, spirituality, and hopeful prosperity, food production started to play a profound role in reshaping human destiny.

In the beginning, food surpluses in good years enabled a much greater degree of role differentiation within farming societies, creating space for less immediately productive roles, such as toolmakers, butchers, and builders. It also created a need for leadership, the foundation of early elites, through grain storage and clearinghouses for surplus food and other goods.

Communities with more decisive leadership and a clear social hierarchy became more productive through community projects such as working irrigation systems, greater security, mediation in disputes, and protection of land and possessions. Farmers traditionally passed food production to the ruling elite in various ways, such as through taxes, levies, or tribute, to support the elite's activities, including building, administration, protection, and/or warfare, as required.

In the Near East, more significant central buildings appeared within villages around 8000 BC. Whether these structures were shared granaries, feasting halls, religious buildings, or the homes of chiefs is still being determined. Indeed, they may have served several of these functions.

The ability to generate and control surplus distribution became a path to power and influence, which, in turn, laid the foundations for the key elements of contemporary economies and cemented our eventual preoccupation with growth, productivity, wealth, and trade—not to mention greater food production!

Early agricultural societies used grain as a form of currency in barter transactions and to pay wages and taxes. One of the world's first currencies was barley grain, used by the Sumerians. The Sumerian civilisation lived in southern Mesopotamia, which is now part of Iraq and Turkey. It was one of the first civilisations to create a flourishing urban civilisation based on agriculture and community life. It expanded irrigation channels to enable year-round farming on an immense scale.

The Sumerian civilisation's first currency, later replaced by a coinage known as the shekel, was based on fixed amounts of barley grain and used as a universal accounting measure for evaluating and exchanging all goods and services. The common denomination was the Sila, equating to roughly one litre of barley by volume.[14]

Cattle were also used as a form of currency in early commodity barter transactions. Consider the word "pecuniary," which means money in several languages today. Its origin is linked to the Latin root "peculium," meaning private property, but it is also associated with the Latin noun for cattle, "pecus." Similarly, the term "capital," when referring to assets, also has roots in cattle; it derives from "capita," meaning "head."

Another food currency used in earlier times was salt. In addition to being used as a food preservative, it retained a high value for bartering transactions due to its relative scarcity. Again, the origins of the word salary come from Roman times when soldiers were sometimes paid in salt. Their monthly allowance was called "salarium"[15] (sal the Latin word for salt).

The Sumerians later invented counting tokens to take stock of these new possessions, such as a record of land, grain or cattle ownership. This was recorded through simple stamps inscribed with pictures impressed in clay tablets representing the objects to be itemised. The same pictograms became more stylised as scribes began drawing them with a wedge-shaped stylus made of reeds. By 3500 BC, the subsequent script that developed, now known as cuneiform, became the world's first written language.[16]

By 2300 BCE, it had become a Sumerian belief that agriculture, animal husbandry, and weaving were introduced to humanity from the sacred mountain Ekur, an assembly of gods, also known as the Anunnaki deities, with parallels similar to those found in Greek mythology and Mount Olympus.

In the earliest Sumerian writings from the post-Akkadian period, around 2154 BCE, the Anunnaki are described as deities in the pantheon, descendants of An, the god of the heavens, and Ki, the goddess of the earth. Their primary function was to decree humanity's fates.

As societies became increasingly unequal following the Agricultural Revolution, religions grew more intense. Roles within society became more restrictive and were rationalised through the conditioning effect of religion. The process of religious evolution spans from a simple form of animism, a belief system in which everything in the natural world, including plants, animals, rocks, and even inanimate objects, is believed to possess a spirit or soul, to polytheism and the belief in multiple deities.

The Neolithic Revolution also paved the way for the ensuing Bronze Age and the Iron Age, with significant advancements in creating tools for farming, warfare, and conquest.

The entire Nile valley was united into the first Egyptian kingdom in 3100 BCE. It's Pharaohs ruled thousands of square kilometres and hundreds of thousands of people.[17]

It was around 2250 BCE when Sargon named himself "Sharru-kin" (Rightful king) and united the Sumerian city-states to forge the first empire of Akkadian.[18] It boasted over a million subjects and a standing army of 5400 soldiers.[19]

The first mega-empires appeared in the Middle East around 1000 BCE: the Assyrians, the Babylonians, and the Persian Empire. In 221 BCE, the Qin dynasty united China. It levied taxes on approximately 40 million people through a complex bureaucracy that employed more than 100 000 officials.[20] The Romans also united the Mediterranean basin around the same time. At its zenith, the Roman Empire collected taxes from up to 100 million subjects and financed a standing army of 250 000–500 000 soldiers.[21]

Before the transition to agriculture, around 10 000 BCE, the Earth was home to approximately 4 million people living as hunter-gatherers in small family groups, a lifestyle they had maintained for the previous 3 million years.

By the beginning of the first century CE, with the transition to agriculture and the creation of food surplus, the world's population had grown to an estimated 190 million people, of which approximately 1–2 million foragers (mainly in Australia, Mesoamerica, and Africa) remained.

The world's first significant Cultural Revolution had begun.

3.3 DEVELOPING NEW FOOD AND TASTE PROFILES

Many historians believe that the origins of brewing beer and wine date back to the Neolithic period, when the first domesticated cereal and fruit crops became available. The earliest evidence of a fermented alcoholic beverage made from fruit, honey and rice was found in China more than 7000 years ago.

Georgia is considered the cradle of winemaking, with a history dating back approximately 6000 years. Early Georgians were making wine when they discovered that grape juice could be turned into wine after burying their kvevris underground.[22] Kvevris are large, egg-shaped clay containers used to create, store, and age Georgian wine. The word kvevri translates from Georgian to "to bury" or "buried." Grapes, skins, stalks, and pips are placed into the kvevris, they are then sealed and left to ferment for five to six months. The underground environment helps to balance the fermentation process, preventing it from overheating.

Fermentation for food processing, as opposed to alcohol production, occurred around the same time, not only because it preserved food but also because it introduced a variety of new flavours and was perceived to improve digestion and have other beneficial effects. Processes for food storage and preservation, such as smoking, drying, and salting meat, as well as fermentation, whether with co-evolved yeasts and lactobacilli (used to produce bread and beer from cereals or cheese from milk), also improved the taste and vitamin intake.

Neolithic farmers stumbled on the practice of domesticating microbes when they tried to preserve food by fermenting it. Fermentation relies on microbes, such as bacteria, yeast, and fungi, to increase the acidity of the food, thereby protecting it against spoilage. Microbes that were proficient in producing palatable and safe fermented products were selected to initiate new batches of the same product, allowing valuable microbes to evolve and be domesticated.

Saccharomyces cerevisiae (*S. cerevisiae*), also known as baker's yeast, has been instrumental in food and beverage fermentation for millennia.[23] Recent genomic evidence suggests that the established beer and bread yeast, *S. cerevisiae*, originated in China before migrating west *via* the route that would later become known as the Silk Road.

Humans also began fermenting milk to create cheeses and yoghurts around 6000 years ago. *Kluyveromyces lactis* (*K. lactis*), also known as milk yeast, is still used in French and Italian cheeses made from unpasteurised milk, as well as in natural fermented dairy drinks like kefir.

The ancestor of this microbe was initially associated with the fruit fly, and it's believed that milk yeast owes its existence to a fly landing in fermenting milk and initiating a liaison with its cousin, *K. marxianus*.

The fly in question was the common fruit fly, Drosophila, which carried the ancestor of *K. lactis*. Although the fly died when it landed in the milk, the yeast had a problem—it could not use the lactose in milk as a food source.

When *K. lactis* arrived with the fly, its cousin *K. marxianus* was already happily growing in the milk. *K. marxianus* could use lactose for growth because it has two extra proteins, which help break down lactose into simple sugars it uses for energy. The cousins reproduced, and the genes necessary for lactose utilization were transferred from *K. marxianus* to *K. lactis*. The result was that *K. lactis* acquired two new genes, enabling it to grow on lactose and survive independently.[24]

On the health side, almost every civilisation regularly ingested fermented milk products for health benefits.[25] The Greek physician Hippocrates is credited with saying what could now be considered prophetic. "All disease begins in the gut." Hippocrates considered fermented milk a food product and a medicine that could cure intestinal disorders.[26]

Pliny the Elder detailed how fermented milk was used to treat gastrointestinal disorders, including gastroenteritis, in ancient Rome. While the term 'probiotic' (bios: life; pro: in favour of) was only coined in the 1960s, various population groups have been using fermented milk products and live bacteria for millennia.

However, milk had to be fermented before it could be ingested. Raw milk, straight from cows, goats, or sheep, was essentially a toxin to adults. Unlike babies and young children, adults could not produce the lactase enzyme required to break down lactose, the primary sugar in milk. "If you're lactose intolerant and you drink half a pint of milk, you're going to be really ill. Explosive diarrhoea—dysentery essentially," says Oliver Craig, an archaeologist at the University of York, UK.[27]

The genetic mutation that enables humans to digest milk, known as lactase persistence, first emerged in Europe approximately 5000 years ago. It likely spread to other parts of the world later. These individuals are said to be lactase persistent, more commonly known as lactose tolerant. This DNA adaptation opened up a rich new source of nutrition, including fats, proteins, and other micronutrients, which were not easily obtainable in cereal crops. Having a readily available substitute for mother's milk also reduced the lactation period and the corresponding infertility period for women, resulting in a greater population for milk-consuming societies.

This two-step milk revolution may have been a key factor in allowing bands of farmers and herders to sweep through central and northern Europe, displacing the hunter-gatherer cultures that had lived there for millennia.[28] The remnants of that pattern are still visible today. In southern Europe, lactase persistence is relatively rare—less than 40 percent in Greece and Turkey. In Britain and Scandinavia, by contrast, more than 90 percent of adults can digest milk.

Long before the popular resurgence of probiotics today, cultures around the globe were experimenting with lactic acid fermentation and other fermented products during the Neolithic era. Vegetables such as cabbages (Sauerkraut), carrots, garlic, soybeans (soya sauce), olives, cucumbers, onions, turnips, radishes, cauliflower and peppers, in addition to fruits such as lemons or berries, offered novel flavours and textures and could easily be replicated by way of home fermentation.

The addition of salt to food was in general use long before recorded history began, and it was used for both seasoning and preservation, helping to eliminate dependence on the seasonal availability of food. It also made it possible to transport food over long distances. As people became increasingly addicted to it, salt became a vital commodity in trade and the economic foundation of several empires and cities during the Middle Ages. The first taxation of salt dates back to 300 BCE when China used a salt tax as the primary source of financing for the Great Wall of China.[29]

Our consumption of salt (40 percent sodium, 60 percent chloride) today is 10 to 20 times greater than 5000 years ago! Because the human body originally evolved to conserve salt from other food sources, it finds it difficult to dispose of this relatively sudden, in evolutionary terms, increase in salt intake.[30] Too much salt in the diet can lead to high blood pressure, heart disease and stroke. It can also cause calcium loss, some of which may be drawn from the bones. As will be seen in a later chapter, most people's kidneys also struggle to keep up with the extra sodium in the blood, which leads to further metabolic complications.

REFERENCES

1. E. Rogosa, *Restoring heritage grains: the culture, biodiversity, resilience, and cuisine of ancient wheats*, Chelsea Green Publishing, White River Junction, Vermont, 2016.
2. H. James Birx, *21St Century Anthropology: A Reference Handbook*, Sage Publications, Thousand Oaks, Calif., 2010.
3. E. V. Koonin and M. Galperin, *Sequence—Evolution—Function*, Springer Science & Business Media, 2013.
4. National Geographic Education, The development of agriculture, https://education.nationalgeographic.org/resource/development-agriculture/.
5. D. J. Cohen, The beginnings of agriculture in China, *Curr. Anthropol.*, 2011, **52**, S273–S293.
6. A. Collins and G. Hancock, *Gobekli Tepe: the Temple of the Watchers and the Discovery of Eden*, Bear & Company, Rochester, 2014.
7. M. Tobolczyk, *The Art of Building at the Dawn of Human Civilization*, Cambridge Scholars Publishing, 2020.
8. M. N. Cohen, G. J. Armelagos and C. S. Larsen, *Paleopathology at the origins of agriculture*, University Press of Florida, Gainesville, 2013.
9. M. R. Sánchez-Villagra, *The process of animal domestication*, Princeton, Princeton University Press, New Jersey, 2022.
10. M. Ferrando-Bernal, Ancient DNA suggests anaemia and low bone mineral density as the cause for porotic hyperostosis in ancient individuals, *Sci. Rep.*, 2023, **13**(1), 6968.
11. J. E. Buikstra, *Ortner's identification of pathological conditions in human skeletal remains*, Academic Press, London, United Kingdom, 2019.
12. P. Ungar, *Evolutions Bite a story of teeth, diet, and human origins*, 2018.
13. A. Mesoudi, Cultural evolution: a review of theory, findings and controversies, *J. Evol. Biol.*, 2015, **43**, 481–497.
14. D. Connors, *A History of Money*, University of Wales Press, 2016.
15. J. Cresswell, *Oxford dictionary of word origins*, Oxford University Press, Cop, New-York, 2010.
16. S. N. Kramer, *Sumerians: Their History, Culture, and Character*, University of Chicago Press, Chicago, 1971.

17. B. J. Kemp, *Ancient Egypt: anatomy of a civilization*, Routledge, Abingdon, Oxon, 2018.
18. J. A. Shoup, *The History of Syria*, Bloomsbury Publishing, USA, 2018.
19. G. Leick, *Historical Dictionary of Mesopotamia*, Scarecrow Press, 2009.
20. Y.-N. Li, *The First Emperor of China*, Routledge, Taylor & Francis, 2018.
21. E. J. Watts, *The Eternal Decline and Fall of Rome*, Oxford University Press, 2023.
22. P. E. McGovern, *Uncorking the Past: the Quest for Wine, Beer, and Other Alcoholic Beverages*, University of California Press, Berkeley, 2009.
23. J. Dalcy, Beer Yeast Is a True International Collaboration, Smithsonian Magazine, 2019.
24. J. Morrissey, A 6000-year-old fruit fly gave the world modern cheeses and yoghurts, The Conversation, 2019.
25. L. Holmes, *Heal Your Gut*, Fair Winds Press, 2016.
26. B. Schindler, *Eat like a Human*, Little, Brown Spark, 2021.
27. A. S. Wiley, *Re-imagining Milk*, Routledge, 2015.
28. H. Brüssow, *Environ. Microbiol.*, 2013, **15**(8), 2154–2161.
29. M. Kurlansky, *Salt*, Vintage, Canada, 2011.
30. G. Stone, *How Not to Die*, Flatiron Books, 2015.

CHAPTER 4

Ceres, the Goddess of Grain and Agriculture

4.1 GRAIN MILLING

Grain milling to produce a staple food existed long before Neolithic times, as did subsequent bread baking from the same grains. Millstones, a saddle stone, quern or handstone, were used to crush early grains such as millet, emmer (a cereal grain closely resembling wheat), and einkorn. As further agrarian settlements evolved, larger horizontal stones, one fixed, the other revolving, known as a quern stone, were powered by hand or animals. The subsequent milled flour was combined with yeast, salt, and water to produce what would have been, in effect, a wholesome wholegrain bread.

Today, world cereal production has reached approximately 2.84 million tons per annum (2024). Roughly 1.2 million tons of maize, 789 million tons of wheat and 538 million tons of rice.[1] Just 55 percent of the world's crop calories are eaten directly by people. Another 36 percent is used for animal feed. And the remaining 9 percent goes toward biofuels and other industrial uses.[2] Cereal grains still contribute around 60 percent of the total calories in developing countries' diets. Maize and wheat comprise nearly two-thirds of the world's food energy intake.[3]

Food and Us: The incredible story of how food shapes humanity
By Seamus Higgins
© Seamus Higgins 2025
Published by the Royal Society of Chemistry, www.rsc.org

The word cereal is named after the Roman god Ceres, the goddess of agriculture, grain crops and fertility; the word also has a linguistic meaning: "to satiate, to feed". Ceres is modelled on the goddess Demeter, the grain goddess in Greek mythology. In the broader sense, Demeter was akin to Gaia, the ancient Greek personification of the Earth.

At the beginning of the first Century CE, of the eight grains commonly used as human food today, six of them, maize, sorghum, rice, oats, rye and millets, were either unknown or barely grown in the Mediterranean world of Roman times.[4] The two grain crops most used were barley and wheat. For various reasons, not least those related to the production of leavened bread, wheat became the more popular grain source for most of the population.

It was recorded at the time that the average grain consumption was about three- and one-third modii per month per person, a modius being a Roman unit of measure for dry goods equivalent to approximately 8.73 litres. Rome needed about 40 million modii a year, or more than a quarter of a million tons of grain per annum, to support a population of approximately one million people.[5] Wheat yields from the thin soil of decomposed limestone, typical of much of the local region, would have been low; therefore, most of Rome's grain, with a higher protein content more suitable for bread baking, was imported from areas as far as Sicily and the Nile Valley.

Now think of the logistics of moving that grain by wooden shipping vessels, storing it at shipping and landing ports, and cart delivery to final distribution centres. Remember that grain is an organic, living product, and even after harvesting, it still needs to breathe.

When stored or shipped post-harvest, grain absorbs oxygen and releases heat, carbon dioxide, and moisture. If this process is not adequately monitored and controlled, the grain will germinate, or bacteria in the air will become active in the presence of the grain, encouraging the growth of mould and fungi, and ultimately causing the grain to rot. Another storage/shipping problem is infestation by insects, such as weevils, grain borers, moths, or beetles, as well as rodents like mice or rats.

Then, as now, in most countries around the world, the price of a staple grain to feed the general populace; however, it was handled and or sourced, was of significant concern to the politicians of the time. Likewise, the interests of the five "corpora" of grain shippers, similar to our present-day quartet of the "ABCDs of grain trading", more details in a later chapter, often threatened to withdraw their services to the city of Rome if their price and or trading conditions were not met!

Vitruvius, a Roman engineer and military architect, detailed the first undershot vertical water wheel to power a set of horizontal millstones using a 90-degree gear mechanism for grain milling around 25 BCE.[7] It was the world's first food processing system purposely designed to operate without being powered by humans or working animals, and was explicitly intended for easier flour and food production.[6]

When Leonardo Da Vinci later immortalised Vitruvius in the 15th Century, combining ideas about art and architecture, human anatomy and symmetry in one distinct and commanding image, entitled "Vitruvian Man", the same Vitruvius could also have been celebrated as the world's first food process engineer (Figure 4.1).

Wheat, as we know it today, 95 percent of all grain currently grown (Triticum aestivum), eventually became more common and, because of its high gluten structure, was quickly adopted by the Romans as a better grain for bread production. The word "flour" is also derived from the Latin word "flos", meaning "flower", or more figuratively, the best part or kind of something.

At the time, barley was considered a staple food for the poor. Indeed, a notable irony of the time was that as Roman citizens became more prosperous, they demanded more white flour products for their tables. Yet a common nickname for their physically fit and robust gladiators, who had to fight each other to the death every other week, was referred to by the historian, Pliny the Elder, as "hordearii," which means "barley men" or "barley eaters".[7]

Due to the demand for white flour, Romans also processed wheat grain to separate the coarser particles, including bran and wheat germ, using woven baskets made from hair or papyrus to sift out a "whiter" flour. Romans were also the first to sieve flour through expensive linen cloth in pursuit of the same "whiter" flour quest.

To this day, Romans and Italians still love their ultra-fine sifted white flour milled to a particle size of 100 microns,

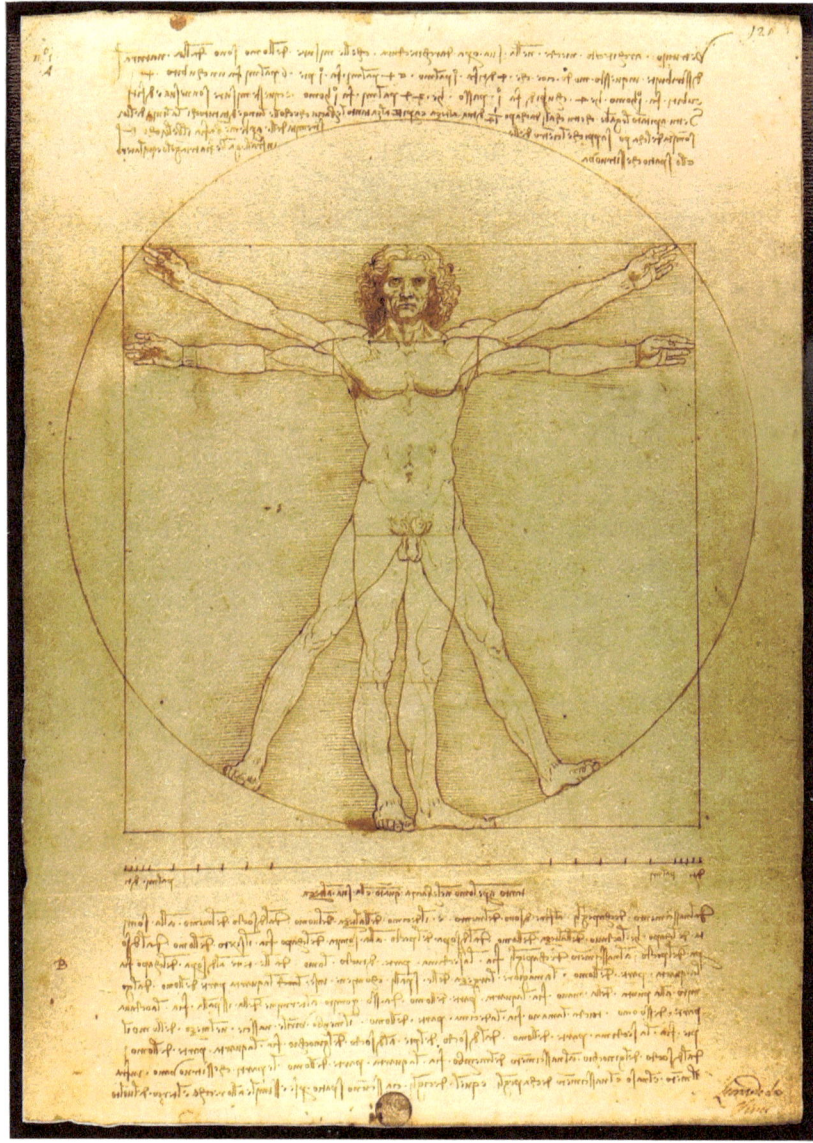

Figure 4.1 Vitruvian Man. Figure in Public Domain. (The drawing '*Homo vitruviano*', showing a man of "ideal proportions", is in the collection of the Gallerie dell'Accademia in Venice).

referred to as "00" or "doppio zero". No self-respecting Italian cook would dream of making fresh egg pasta without their "00" flour.

With the expansion of the Roman Empire into Europe and Britain, its milling technology also followed. By the middle of the 11th century, Britain's Domesday Book (1086) lists more than 5000 mill sites in the UK. By the end of the 13th century, this number had more than doubled.[8]

At that time, Feudalism was in full bloom for most of northern Europe, with its heartland comprising the rich agricultural lands of the Seine Valley in France and the Thames Valley in England. As King Alfred the Great summarised at the time, the (medieval) population was divided into three groups: 'those who pray' (clergy), 'those who fight' (knights, soldiers, aristocrats), and 'those who work' (peasants).[9]

Milled barley was still primarily used as a poor man's potage or for brewing ale, a typical daily drink instead of water. In 1324, King Edward II of England used three grains of barley as the model for the size of an inch and standardized the measurement as "three grains of barley, dry and round, placed end to end lengthwise."[10]

Then, as now, for all food manufacturers, trust and reputation were significant issues for the miller. The Norman Conquest in the 11th century introduced a feudal system to the UK, and 'soke rights' forced everyone to have their grain milled at the mill owned by their Manorial Lord. The miller, often under pressure from the local lord for better returns, would tamper with the flour. Some of the earliest forms of food contaminants, commonly used to bulk up flour, included sand, ash, sawdust and mustard flour. Ground animal bones were also used as an additive to make flour look whiter.

The "art" of grain milling has also given us some rich metaphors and sayings in English.[11]

"*First come, first served*" was the law for many millers in many countries. It could take days for a farmer to have his grist ground, and the law was designed to prevent impatient customers or friends of the miller from queue jumping.

"*The daily grind*"; the repetitive nature of milling and the concept of the daily (same old!) grind.

"*The run of the mill*" refers to the same day-to-day ordinary grind.

"*Fair to middling*"; the quality of ground meal would be fair, middling, or fine. To be "fair to middling" is to be below one's best.

"Keeping your nose to the grindstone"; If the mill stones were set incorrectly, they could grind too hot, and the flour would become cooked, emitting a burning smell, and or worse, the flour could burst into flames. Hence, the warning to keep one's sense of smell around the grindstones because most mills were made of wood and could, and frequently did, burn to the ground.

"Rule of Thumb": to test the quality and mill grind of the flour, the miller would take a pinch of flour between his thumb and forefinger.

Some of these colloquial sayings of the time were anything but complimentary to the miller, such as:

"The Miller's Golden Thumb" is most probably a play on the miller's "rule of thumb." It was said that when weighing finished flour products, the miller always had a knack for holding his thumb on the right side of the weighing balance.

"As stout as a miller's waistcoat, that takes a thief by the throat every day" – Old German saying.

My favourite one is an Epitaph on an Essex churchyard of a miller named Strange; *"Here lies an honest miller, and that is Strange –"*

Post-Norman times, when common trades were adopted as people's surnames, to this day, the trade-inspired surname "Miller" is still the 7th most popular surname in the US.[12] In Europe, other forms of "Miller", Mueller, Muller, Moeller, Muller (German), Moulin, Moulinier, Meunier, Molyneaux (French), Molnar (Hungarian), Melnikov (Russian), Farina, Molinari (Italian), Molander (Swedish), and so on.

It was in 1369 when a group of French millers who owned and shared a perpetual lease on the Garonne River in Toulouse formed the Société des Moulins de Bazacle. Thus creating the world's first limited liability company registered as its own legal entity distinct from its shareholders.[13] At the time, the mill owners signed a profit-sharing agreement. They issued 96 Société des Moulins du Bazacle shares that would trade at a value dependent on the mills' profitability. The original stock offering was underwritten by a group of local seigneurs who shared the profits according to the number of shares they possessed. The shares of the society eventually came to be traded on the open market in Toulouse, and their value fluctuated according to the mills' profitability. Capitalism born from food supply was now ready to expand and evolve beyond just food.

Just as barley grain had become the world's first currency, wheat milling created the world's first limited public company. It was 230 years later, in 1600, when Queen Elizabeth I granted a charter to a broad group of London merchants to establish the East India Company. Soon after, in 1602, the Dutch East India Company issued shares that were made tradable on the Amsterdam Stock Exchange.

By the 16th century, water and wind power had become the most important source of motive power for milling grain in Britain and Europe. Water and windmills peaked at more than 20 000 mills in the UK by the 19th century.

In 1860, a Swiss-educated, German mechanical engineer named Henry Simon arrived in Manchester, UK. An accomplished engineer, he founded his own business in 1878. He built the first milling plant in Manchester based on what was referred to as a 'gradual Reduction' milling system using steel rollers instead of stone.[14] In 1881, he built the world's first automatic belt-driven steam-powered roller flour mill for McDougall Brothers, a predecessor of Rank Hovis. It was in 1890 when Adolf Buhler of Uzwil, Switzerland, delivered his first Buhler wheat mill. Miag of Germany followed in the 1920s, and Golfetto followed in Italy.

In 1919, George T Smith applied for a patent for his milling equipment design in America, and the term "patent" flour produced by his equipment was first coined. This enabled the miller to advertise his flour as being produced from a "patented" process.

So, after almost two millennia, the Roman quest to produce pure white flour was achieved by way of the gradual reduction milling process. Within the space of just three decades, at the turn of the 20th century, the art of grain milling had finally evolved to gain maximum extraction of the wheat kernel's white endosperm (a starchy carbohydrate) without the "contamination" of the grain's darker components such as its wheat germ, the aleurone layer or its outer fibrous bran husk.

The benefits of this new process soon gained rapid acclaim within the milling industry; not only could millers now produce a good-quality white flour that sold at a premium price relative to traditional stone-milled flour, but they also successfully removed the fatty wheat germ component (the grain's embryo), which constituted 3–4 percent of the wheat kernel, to increase their white flour's shelf life from approximately 6 weeks to 3–6 months.

Typical mill extraction rates of 70–75 percent of white flour could now be achieved, depending on the grain quality milled. New markets were also developing due to animal feed requirements for all the so-called wheat "offal" generated.

In 1878 the President of the newly formed National Association of British and Irish millers, Mr Simon Hadley, outlined the general aims of this new industry association; 'The collection of information bearing upon all departments of the trade, technical, practical, and commercial, to improve the quality of its products and increase the ratio of its profits'.

Rice milling in Asia also followed a similar path. Today, rice is the staple food for more than half the world's population, including an estimated 640 million undernourished people in Asia. It is also rapidly becoming a food staple in Latin America and Africa. More than 90 percent of the world's rice is grown in Asia, principally in China, India, Indonesia, and Bangladesh. Smaller amounts are produced in Japan, Pakistan, and various Southeast Asian nations as well as Europe, North and South America, and Australia.

In Chinese logographic writing, the word "eat" means "eat rice". And just as milled rice grains have been brown for millennia, with its outer inedible hull removed, whole grain rice retained its bran and germ layer, constituting the brown or tan colour of "brown "or wholegrain rice.

Indeed, the Chinese character for vitality, or "Jing" (精), is composed of the characters "米" (rice) and "青" (plant or vegetables). The Chinese character for energy, or "Chi" (氣), is made up of the characters "气" (breath/breathing) and "米" (rice). Early Chinese civilisation obviously believed that vitality and energy depended on proper breathing, a plant-based diet, and brown rice.

That philosophy changed around the end of the 19th century with the advent of more sophisticated rice milling equipment, similar to wheat milling's 'gradual reduction' mills, which enabled the mechanical separation of rice bran and its germ layer from the more traditional brown rice to produce a white, polished grain.

At the same time, a relatively new debilitating disease called Beriberi occurred. Symptoms included difficulty walking.

Loss of feeling (sensation) in hands and feet. Loss of muscle function or paralysis of the lower legs, including mental confusion and speech difficulties. Unknown at the time, Beriberi was the classical manifestation of a chronic dietary deficiency in vitamin B1 (Thiamine).[15]

By removing the rice bran to produce a white, polished grain, the B1 vitamin, Thiamine, content of the grain was diminished by 90 percent. Beriberi became a significant health problem in East Asia and other regions, where polished rice had become the primary staple food.

In 1897, Christiaan Eijkman, working in the Dutch East Indies (now Indonesia), found that a Beriberi-like disease could be produced in chickens by feeding them a diet of polished rice. It was not until 1912 that Casimir Funk, later known as the "father of vitamins," demonstrated that Beriberi-like symptoms induced in pigeons could be cured by feeding them white rice supplemented with a concentrate made from rice polishings.[16]

Following his discovery, he proposed that Beriberi and several other conditions were due to diets deficient in specific factors that he called "vitamines,". He chose the word "vitamine" as derived from the Latin "vita," meaning life, and "amine" because he believed vitamins contained the protein of amino acids. Funk coined the term "vitamin" to describe compounds that were "vital" to health and centered on an "amine" group. He also postulated the existence of vitamins B1, B2, C, and D. He was the first to isolate the vitamin niacin (vitamin B3) in 1912.

Today, as in earlier times, when rice is harvested, the inedible hull is removed to produce whole-grain brown rice. Brown rice is a highly nutritious, gluten-free grain with impressive vitamins, minerals, and beneficial compounds. It is an excellent source of fiber, magnesium, phosphorus, vitamin B1, vitamin B3, vitamin B6, potassium, selenium, and calcium. Consuming whole grains like brown rice can help prevent or improve several health conditions, including diabetes and heart disease.[17]

Parboiled rice, sometimes called converted rice, originated in ancient India and is now commercially applied to more than 25 percent of the world's rice supply. It involves soaking, steaming, and drying the rice while still in its husk after harvest but before milling. In contrast to other raw rice, parboiled rice appears glassy and translucent, with a light amber colour before cooking. After cooking, it is firmer, fluffier, and less sticky.

Although parboiling decreases the thiamine content compared to brown rice, milled parboiled rice still contains more thiamine than milled raw rice, even with the same degree of milling. This feature of par-boiled rice can be explained by an inward diffusion of vitamins to the endosperm during the parboiling process. The milling and polishing process of white rice removes most of its essential vitamins and other nutrients. In one sense, polished rice is nothing more than just a refined starch.

Just as per the rise of Beriberi in Asia, post the advent of new grain milling processes that remove the outer bran layer and the grain embryo of maize/corn, was also instrumental in introducing a new disease to the world, namely Pellagra. The disease name was first coined by Dr Frapolli in 1771 from the Italian 'pelle' for skin and 'agra' for rough, thus describing the disease's most striking feature. At the time, nobody realised that Niacin, vitamin B3, or lack thereof, was the cause of Pellagra. The nutrition-related disease was characterised by what was known as the 4 D's: dermatitis, diarrhoea, dementia, and, in untreated cases, death.[18]

During the early 1900s, pellagra deaths were recorded in every state in the US, but the disease was much more pronounced in the South. Maize meal, or corn grits as it is more commonly known, had only recently become a popular foodstuff. As "King Cotton" and textile mills dominated the South's post-Civil War economy, many families converted all their farmland to cotton. They stopped planting vegetables and keeping livestock. As a result, many poor Southerners now ate almost exclusively what was called the three Ms: low-quality meat (salted pork), molasses and meal (industrially refined cornmeal)—the same cheap gruel was also served at orphanages, asylums and other state-run institutions.

Working with the same institutions, Dr Joseph Goldberger saw a flaw in the fly-infection theory posited for pellagra at the time. Institutions with pellagra cases consistently reported that staff members, presumably vulnerable to the same flies, were not diagnosed with the disease. The recent identification of diseases caused by dietary deficiencies further convinced Goldberger that a sole corn-based diet could be the cause of Pellagra.

Goldberger focused on identifying the missing dietary element, which he called the P–P factor, for pellagra prevention.

In 1922, he attempted to induce black-tongue disease—the canine equivalent of Pellagra—in his laboratory dogs by feeding them a diet typical of poor Southerners, supplemented with brewer's yeast to stimulate their appetite.

The dogs remained healthy, prompting suspicion. Without the yeast, the dogs developed Pellagra. Repeated testing on the dogs and human subjects confirmed that brewer's yeast, rich in Niacin! contained the P–P factor that cured and prevented Pellagra. It was only in 1937 that work by Dr Conrad Elvehjem, inspired by Goldberger, isolated the factor that cured Pellagra. This factor proved to be the same nicotinic acid, Niacin, discovered by Casimir Funk in 1912.

Between 1907 and the early 1940s, Pellagra affected over 3 million individuals in the South, with more than 100 000 deaths recorded. Today, Pellagra is mainly relegated to history lessons and medical reference books. Although Pellagra still presents problems in places where milled corn meal still underscores the daily diet. Epidemiologists have recently recorded the disease in Malawi, Mozambique, Angola, Zimbabwe, Nepal, and Angola.

Another food irony of the times, with milled corn and Pellagra in the US, is the fact that dent corn (Zea Mays), as will be seen later in the book, originates from Mesoamerica! The earliest undisputed domesticated maize cobs are from a Guerrero, Mexico cave, 4280-4210 BC. The term "maize", as used for corn, is derived from the ancient word "mahiz" from the now-extinct Taino language of the indigenous people of pre-Colombian America and was a staple food for native Americans for centuries.[19]

However, people across Mexico and Central America use a traditional method known as nixtamalisation to process their maize. The Aztecs discovered that wood ash from cooking fires mixed with water created an alkaline lime mixture that softened and partially dissolved the kernel's tough outer skin, making corn easier to cook and to make into masa (dough).

In addition to altering the smell, flavour and colour of the maize, nixtamalisation also provides several nutritional benefits, including increased bioavailability of the vitamin B3, Niacin, increased calcium intake due to its absorption by the kernels during the steeping process and increased resistant starch content, which serves as a source of dietary fiber.

Nixtamalisation also significantly deactivates (90–94 percent) of the mycotoxins produced by moulds that commonly infect raw maize. This ancient technique of nixtamalisation is still used today. The process is employed using both traditional and industrial methods, particularly in the production of tortillas and tortilla chips (excluding corn chips), tamales, hominy, and many other items.

With the expansion of international trade, corn and corn milling have become widespread globally, making corn a significant food source in Africa, Europe, and the United States. People boil and eat it whole, mill it to make flour and cereals, and cook it in sweetened milk for dessert. Corn can also be processed for various uses, including sweeteners (such as corn syrup), alcohol (such as whiskey), cooking oil, and bioethanol as a fuel for motor vehicles. Humans now get 19.5 percent of their calorie intake from corn or corn products.

REFERENCES

1. Food and Agricultural Organisation; FAO cereal supply and demand brief, https://www.fao.org/worldfoodsituation/csdb/en.
2. B. Plummer, How much of the world's cropland is actually used to grow food? https://www.vox.com/2014/8/21/6053187/cropland-map-food-fuel-animal-feed.
3. G. A. Nayik, T. Tufail, F. M. Anjum and M. J. Ansari, *Cereal Grains*, CRC Press, 2023.
4. R. Laurence, *Roman Passions*, Bloomsbury Publishing, 2010.
5. G. Rickman, *The Corn Supply of Ancient Rome*, Clarendon Press, Oxford University Press, Oxford, New York, 1980.
6. C. Rossi, F. Russo and F. Russo, *Ancient Engineers & Inventions*, Springer Netherlands, Dordrecht, 2009.
7. Pliny The Elder, The Natural History of Pliny, BoD – Books on Demand, 2023.
8. E. King, *Medieval England*, Tempus Publishing, Limited, 2001.
9. R. Abels, *Alfred the Great*, Routledge, 2013.
10. C. Marlowe and M. R. Martin, *Edward the Second*, Broadview Press, Peterborough, Ont., 2010.
11. T. Hall, The Mapperley Story - Ilkeston Advertiser, 1969.

12. P. Nightingale, *Trade, Money, and Power in Medieval England*, Taylor & Francis, 2023.
13. Yale School of Management, The fascinating 600-year history of a French mill, the world's oldest shareholding company Aug. 19th 2014.
14. H. Simon, *Our Foundation*, https://www.henrysimonmilling.com/aboutus/our-foundation.
15. D. Lonsdale and C. Marrs, *Thiamine Deficiency Disease, Dysautonomia, and High Calorie Malnutrition*, Academic Press, 2017.
16. A. H. Ensminger, *Foods & nutrition encyclopaedia*, CRC Press, Boca Raton, 1994.
17. B. O. Juliano, *Rice in human nutrition*, Fao, Rome, 1993.
18. A. M. Kraut, *Goldberger's War*, Hill and Wang, 2021.
19. T. Standage, *An edible history of humanity*, Walker & Co, New York, 2010.

CHAPTER 5

"Give Us, This Day, Our Daily Bread"

The earliest evidence of flatbread baked by hunter-gatherers was found in Jordan and dates back approximately 14 400 years, predating the advent of agriculture by at least 4000 years.[1] Indeed, the findings suggest that bread production based on wild cereals may have also been instrumental in encouraging hunter-gatherers to begin cultivating cereals.

As the Neolithic Revolution developed and cereal growing became more widespread worldwide, bread became a staple food for people of all socioeconomic statuses. It even evolved its own mystic and religious relevance. Previous religious beliefs, connected to the animal world through Animism, were replaced with a new mythology and ritual representation, which tended more towards plant life, with wheat and bread symbolising the Earth's fecundity.

In the Sumerian language, the world's oldest known written language, the word for bread is Nin-da. Nin means "lady, woman, or goddess".[2] Since all food comes from the Earth, it came to be regarded as a generous Mother Goddess who nourishes her offspring, human or otherwise. Demeter was the Olympian goddess of agriculture, grain and bread who sustained humanity with the Earth's rich bounty. She was depicted as a mature woman, often wearing a crown and bearing sheaves of wheat or a cornucopia (horn of plenty).

The Egyptians called bread "Aish baladi". Baladi means traditional or authentic in English, but "Aish" means life, which is how Egyptians have perceived bread since ancient times.[3] At that point in history, bread would also have been understood as a nutritional necessity. As religion moved from deity worship to monotheism, the religious significance of bread, often referred to by authors and poets as "the staff of life," was carried into Christianity and continues to this day undiminished.

Bread is one of the most potent symbols in the Christian faith. It is mentioned at least 492 times in the Bible, from Genesis through Revelation. It has a variety of meanings and symbolism.[4] "Then Jesus declared, "I am the bread of life. Whoever comes to me will never go hungry, and whoever believes in me will never be thirsty." – John 6:35

One of the seven sacraments of the Catholic Church, the Eucharist is a ritual in which, according to Catholic theology, unleavened bread, the host, and wine blessed by a priest becomes the body, blood, soul and divinity of Jesus Christ. The Jewish religion still uses unleavened bread to celebrate Passover and their rescue from Egypt's pharaoh. Within the Hebrew Bible, a prophetic book called Micah, thought to be written around 722 BCE, prophesies that a messiah would come from Bethlehem, meaning the 'House of Bread' in Hebrew and Aramaic.[5]

In Roman times, the average inhabitant consumed at least 70 percent of their daily calories in the form of cereals, such as bread, pulses (pottage), and legumes. As mentioned previously, bread was an essential part of the Roman diet, with more well-to-do people eating wheat bread and poorer people eating bread made from barley. The types of bread made would have been both leavened and unleavened flat bread. As mentioned in Virgil's Aeneid,[6] when Aeneas' son Ascanius exclaims: "Look! We've even eaten our plates!" Much later, around the 18th century, people from Naples covered the same flatbreads with savoury toppings and "invented" the Pizza.

As with milling technology, the Roman way of life soon spread throughout Europe after the Roman conquest. The importance of milling and bread in the early medieval period is highlighted by the old English words for lord (hlaford) and lady (hloefdige), meaning loaf keeper and loaf kneader, respectively.[1]

In the Middle Ages, bread became a staple food for most people. Mantou is made in China by steaming or frying wheat dough. Tortillas are made from cornmeal, as was the tradition among the Aztecs in Mesoamerica, and German and Scandinavian black bread is made from Rye. Indeed, German bread consists of many types of rolls, black bread and their world-famous pretzels, which are salted and flavoured with cumin or poppy seeds. Pretzels, or pretiolas as they were first known, had their humble beginnings around 610 CE and became the world's first and oldest snack food.

Most breads in India are unleavened and are usually fried. Naan is an Indian flatbread traditionally baked in a clay tandoori oven. The dough is placed on the sides of the oven and develops a teardrop shape as it stretches while it hangs in the oven. Writing from Baghdad in the 10th century, the author of the first Arabic cookbook, Ibn Sayyar al-Warraq, gives his thoughts on the best kinds of bread to eat:[7] "Wheat bread agrees with almost everybody, particularly varieties made with a generous amount of yeast and salt and allowed to fully ferment and bake well. Such breads are lighter and digest faster".

As in Roman times, bread was typically eaten with butter or dripping during the Middle Ages, sometimes accompanied by more substantial proteins such as cheese, beans, or meat. Refined white flour was traditionally reserved for special occasions, such as feast days, when spiced bread or cakes were made.

In medieval Europe, bread was a staple and became an integral part of the food serving arrangement. A standard table setting of the day was set with what was known as a trencher, a piece of stale bread roughly the size of a modern plate. This piece of bread also served as an absorbent base for stews and gravy.

The first food law in British history to regulate the production and sale of food began in 1266 CE. The "Assisa Panis et Cervisiæ", or Assize of Bread and Ale, was developed in late Medieval English Law to regulate the price, weight and quality of manufactured bread and beer.[8] The law specifically focused on the quality, weight and cost of bread and beer. As bread prices were dictated by the price of milled flour and fines for overcharging were high, bakers regularly gave away extra loaves to ensure they complied; hence the expression 'a baker's dozen' being thirteen loaves.

The French, of course, have also always been into bread! One of the oldest descriptions of leavened bread comes from the Roman philosopher Pliny the Elder when he describes how the Gauls used foam from their beer to bake what he called "a lighter kind of bread than other peoples."

Obviously, the Gauls had discovered sourdough and had learned you could hold back a portion of yeasty dough and add it to the next day's bake, enabling a more consistent rise. Using beer froth would have introduced another bacterium into their yeast starter, one like Lactobacillus, which, instead of alcohol, produces lactic acid. Although it creates a slightly sourer taste, hence the sourdough name, the same bacterium also helps the bread last longer without spoiling as most microbes can't handle the acidic environment created by the Lactobacillus.

When we think of French bread, the French stick, better known as the baguette, usually comes to mind. It is a long, thin, crusty loaf that literally translates from the French words for a stick. Outside of France, the baguette is often considered a symbol of French culture. Made from French wheat, which has a lower protein content than other global kinds of grain, the French have made long, thin bread since the mid-18th century.

Some say Napoleon Bonaparte created the French baguette to allow soldiers to carry bread more easily. Since the round shape of other breads took up a lot of space, Bonaparte requested they be made into skinny stick shapes with specific measurements to slide into the pants of his soldiers' uniform.[9]

It has also been said that the lack of quality bread and the rise of the "Age of Enlightenment" led to the French Revolution. The National Assembly would win the revolution with the death of the monarchy (King Louis) in 1793. Afterwards, a set of regulations were made to eliminate social inequality. A translation of the legislation regarding bread is detailed below.

Richness and poverty must both be eliminated from the government of equality. It will no longer make bread of wheat for the rich and bread of bran for the poor. Under penalty of imprisonment, all bakers will be required to make only one type of bread: The Bread of Equality.

The French are very proud of their "equality" bread! On January 13, 2018, it was reported that French President Emmanuel Macron supported a French baker's call for the baguette to be recognised by the United Nations as a global "cultural treasure." He claimed that French bread "was the envy of the world."

To protect their sacred baguette and or baking traditions, the French government passed a new law in 1993 called the Décret Pain. The law states that traditional baguettes must be made on the premises where they're sold and can only be made with four ingredients: wheat flour, water, salt and yeast. They can't be frozen at any stage or contain additives or preservatives – which also results in them going stale within 24 hours![10]

As enshrined in the above French bread law, bread has been made worldwide for centuries using the same four ingredients. Indeed, before the advent of the new steam grain mills, as described earlier, the flour used for all breads would have been stone-ground, whole-grain wheat, barley, Rye, or corn. Coinciding with the introduction of new flour milling systems, and refinements in microbiology following the work of Louis Pasteur in 1879, also led to more advanced methods of culturing pure strains of baker's yeast.

Great Britain introduced specialised growing vats for the production of *S. cerevisiae*. Around the turn of the century, centrifuges were used in the United States to concentrate the yeast, turning yeast production into a new industrial endeavour. Unlike most of Europe, which managed to retain its local and more traditional baking methods the era of industrial bread baking had begun.

In America, between 1890 and 1930, at the beginning of this period, bread was the country's single most important food, and 90 percent was baked in homes. By the end of the same period, bread was still the country's number one food, but 94 percent of it was baked by commercial bakeries outside the home.[11]

Machines were now replacing human hands to bake bread. Alexander Taggart, a third-generation baker from the Isle of Man, founded the United States Baking Company, which later merged into the National Biscuit Company, now known as Nabisco. He declared that their newly branded, industrially produced "Wonder Bread" was the future, unlike anything that could be baked at home. As Clutter Magazine commented at the time, this new, virgin white, 1.5-pound loaf perfectly evoked the otherworldliness of the enormous manufacturing system that was seen as America's future.

By the mid-thirties, 80 percent of commercially produced bread in America was wrapped and sliced. Hence, the expression "the best thing since sliced bread" refers to the convenience of the same invention.[12] Britain followed a similar path with industrial white bread production, with one notable exception being their Hovis loaf.

In 1886, Richard "Stoney" Smith, a miller from Stone in the UK, perfected a new method of steam-cooking flour that preserved wheat germ for bread baking without destroying its nutrients. His patented process produced a new kind of bread that had three times the natural germ content but without the grittiness associated with other whole-meal breads at the time.

In 1890, a national competition was held to replace the somewhat clumsy "Smith's Patent Process Germ Flour" name, and the winner was offered a prize of £25. The winner, Herbert Grime, a Latin teacher, suggested the name "Hovis", from the Latin "Hominis Vis", meaning "Strength of Man". In 1896, the Hovis bread company received a royal warrant as bread suppliers to Her Majesty the Queen.

During World War 2, Britain, importing more than 60 percent of its wheat and flour supply from Canada, introduced new regulations for the milling of bread flour in an attempt to reduce shipping convoys crossing the Atlantic. In 1942, all flour millers were instructed to increase flour extraction from wheat from around 70 percent to 85 percent. All ensuing bread bakers were banned from producing any type of bread other than the newly regulated "National Loaf".

Sarcastically nicknamed "Hitler's Secret" by the British population who were used to eating whiter bread, the same product was not well received. The high fibre dense flour created a loaf which, although nutritious, was heavy and beige/ greyish in colour, crumbly in texture and stale by the time it was purchased.[13]

Bakers were only allowed to revert to white bread manufacturing in 1950, and the baking and production of the National Loaf ended in 1956. After the war, the UK government partnered with the baking sector to establish the British Baking Industries Research Association (BBIRA) in the Hertfordshire village of Chorleywood. The partnership was designed to strengthen and protect the British baking industry and local wheat farmers.

The work of the scientists at Chorleywood in 1961 led to a new way of producing white bread. This made the average loaf in Britain 40 percent softer, reduced its production cost, and doubled its shelf life.

This new method of bread production, "The Chorleywood Bread Process' (CBP), is achieved by a process of energy-intensive batch mixing without bulk fermentation. By adding hard fats, extra yeast, and certain chemicals, and mixing them at high speed, you can prepare the dough to bake in a fraction of the time it usually takes. However, additional inputs, such as chemical oxidising agents, emulsifiers, and amylase, are required, along with extra volumes of refrigerated water, to cool the dough heated by the energy-intensive mixing process.

Indeed, I recall a bakery manager colleague in the late nineties advising me that the secret to making good bread in the UK is to make water stand up straight and look white! This statement is not without foundation as the CBP adds more water to the dough mix, typically an additional one to four percent. Emulsifier additions such as E481, E472e, flour treatment agents, ascorbic acid (Vitamin C) and sodium steroyl lactate are also added to the mix. Other additions include soy flour to improve the dough's handling and machineability, to give the bread better volume and crumb softness. Extra salt is also used to "tighten" the bread and compensate for the loss of flavour and shortened fermentation time. Preservatives such as calcium propionate or sorbic acid are also added to extend shelf life.

Perhaps what is least known about this method beyond the bakery trade is the extensive use of enzymes and amylase in the dough mix. Because these added products are considered processing aids and are "used up" in the bread-making process, there is no legal requirement to include or display these additives on the bread product's ingredient list.

It would be another forty years before Carlos Monteiro and his team developed the NOVA classification and the term "ultra" processed foods to categorise foods based on their level of processing. As will be seen later, the CBP bread-making process can undoubtedly be considered one of the first "ultra-processed" foods. The CBP process is now used in more than 30 countries worldwide, including Australia, South Africa, South America,

Turkey and even on supermarket shelves in France. Over 80 percent of all loaves in Britain are now made the Chorleywood way.

Members of the Real Bread Campaign in the UK argue that as we look ahead, we must ask what we value in our daily bread. Is it its height or squishiness, as the (CBP) marketers might have us believe? Alternatively, is it the reduction of time and financial cost of production that BBIRA sought? Is assessing the value of bread about embracing time, flavour, community, and craft, or is it all about price?

The Real Bread Campaign believes that a future for bread in Britain that embraces its true values and conserves resources is far more likely to protect and preserve the British people's interests, health, and culture rather than using improvers, emulsifiers, and preservatives.[14]

One wonders if Ceres or Demeter could talk what their thoughts would be on how bread baking techniques have evolved over the past two millennia.

REFERENCES

1. W. Rubel, *Bread*, Reaktion Books, 2011.
2. S. N. Kramer, *Sumerians: Their History, Culture, and Character*, University of Chicago Press, Chicago, 1971.
3. B. Ohara, *Let's Bake Bread!* Artisan, 2023.
4. C. Ingram and J. Shapter, *The Bread Bible*, Southwater, 2017.
5. A. Ferreiro and T. C. Oden, *The Twelve Prophets*, Intervarsity Press, Downers Grove, IL, 2003.
6. Virgil, *The Aeneid*, Signet Book, 1961.
7. N. Nasrallah, Annals of the Caliphs' Kitchens Ibn Sayyār al-Warrāq's Tenth-Century Baghdadi Cookbook, Boston Brill, 2014 [cited 2019 Sep 27]. Available from: https://brill.com/view/title/12145?lang=en.
8. C. MacMaoláin, *Food law: European, domestic and international frameworks*, Hart Publishing Ltd., Oxford, 2015.
9. Baguette, 2024, October 26, In Wikipedia, https://en.wikipedia.org/wiki/Baguette.
10. D. Leader, *A Slow Rise*, Penguin, 2024.
11. Saddleback Educational Publishing, *America Becomes a World Power 1890–1930*, Saddleback Educational Publishing, 2010.

12. thewere42, The Origin of Bread and the Phrase "The Best Thing Since Sliced Bread", https://thewere42.wordpress.com/2014/05/05/the-origin-of-bread-and-the-phrase-the-best-thing-since-sliced-bread/.
13. Tea toast and Travel, the national loaf, http://www.teatoastandtravel.com/the-national-loaf/.
14. Real Bread Campaign, Chorleywood Process, https://www.sustainweb.org/realbread/pappy_birthday/.

CHAPTER 6

Regional Food Develops

The difference in regional food preferences was passed down through subsequent generations *via* oral communication. By the Middle Ages, people consumed a variety of fermented foods and drinks, depending on the raw materials, environmental conditions, seasonality, social class, and local taste preferences.[1]

These taste preferences primarily resulted from how well each community adapted to what was available or could be locally sourced and produced. The humid tropics never developed prosciutto ham or hard-cured cheese, as the conditions for making these products do not exist in such a climate. Ecological conditions, and later religion, in a broader sense, determined what people in different places could eat and when, and they made a virtue out of necessity by adapting to what was available.

If one were living in Northern Europe, close to the sea, such as the Netherlands or Scandinavia, one's diet would likely comprise herring and cheese. You would like caribou, seal meat, and whale fat in the Arctic Circle, as the Inuit do. In Mesoamerica, the diet consisted of three staple foods: maize, beans and squash, to which tomatoes, chilli and salt were added. Meat would have included domesticated animals such as dogs, turkeys, and ducks.

Aboriginal Australians, still living as hunter-gatherers, foraged off the land. Their diets relied on nutrient-rich and high-fibre ingredients, including native herbs, spices, fruits, seeds, and

---

Food and Us: The incredible story of how food shapes humanity
By Seamus Higgins
© Seamus Higgins 2025
Published by the Royal Society of Chemistry, www.rsc.org

nuts. Meat sources included kangaroo, emu, crocodile, and witchetty grubs.

Life in Ancient China was hard for the general population, as most farmers were poor. Farmers owned chickens, pigs, and sometimes an ox or mule. In the North, people grew crops of wheat or millet, while in the South, they grew rice. People drank wine made from rice or millet. They also drank tea. In Europe, it was wine or beer, depending on the location. In the South, wine was the typical drink for both the rich and the poor alike (though the commoner usually had to settle for cheap, second-pressed wine). At the same time, beer was the commoner's drink in the North, and wine was an expensive import.

Food in the Middle Ages was considered expensive. Slow transportation and food preservation techniques (based on drying, salting, smoking, and pickling) made the long-distance trade of many foods costly. Its preparation and preservation changed little from the fifth century until the end of the 17th century, the so-called Age of Enlightenment.

While cooked food became a norm, what went into a dish and its quality largely depended on the degree of social class.[2] Cooking always involved an open flame, and most cooking was done in stew pots since it was the least wasteful use of cooking juices and firewood.

Medieval society was highly stratified. In a time when famine was commonplace and social hierarchies were often brutally enforced, food was an important marker of social status in a way that has no equivalent today in most developed countries. Society mainly consisted of three estates, commoners—the working classes—the largest group, the clergy, and the nobility.

Medieval diets in the upper echelons of society revolved around showcasing and displaying extravagance. Bread was considered so ordinary that serving it plain was regarded as poor form. Instead, a medieval lord's table would typically consist of various meats, desserts with enriched dough, and a generous selection of imported spices.[3] Aristocratic estates provided the wealthy with freshly killed meat, river fish, and fresh fruit and vegetables. Cooked dishes were heavily flavoured with valuable spices such as caraway, nutmeg, cardamom, ginger and pepper.

Monasticism became increasingly popular in the Middle Ages, as religion dominated European society. Monasteries served as important centres of learning, spirituality, and community service. Monks served the church by copying manuscripts, creating art, educating people, housing travellers, nursing the sick, and assisting the poor.[4] Convents were especially appealing to women. They were the only place they would receive any education, and they also let them escape unwanted arranged marriages.

Monks themselves were, of course, individually poor as they had few possessions of any kind. Still, the monastery itself was one of the wealthiest institutions in the medieval world. Consequently, monks were well-catered in the one area that mattered most to most of the population: food and drink. All monasteries always had a supply of bread, fish, seafood, grains, vegetables, fruit, eggs, cheese, and plenty of wine and ale. Monks typically had one meal a day in winter and two in summer.

In medieval monasteries, food manufacturing, other than milling, also evolved. I use the word manufacture here as per its Roman roots, from the verb to "make", which has a middle French root from the Latin "manus" meaning hand and "facture" meaning "working".

The monks at the monastery of Ruiel en Brie impressed the Roman Holy Emperor in 774 CE with their offering of a soft-salted, fermented cow's milk, sprayed with a fungus (Penicillium) to create a firm crust over a gooey center, and left in a cellar to mature for six weeks.[5] Their iconic white-crusted "Brie" cheese is still a favourite worldwide today.

Other cheese favourites from various monasteries include Parmesan from the Benedictine abbeys in Parma and Roquefort from the Conques Abbey in Aveyron.

As the Catholic Church, dominant in Europe during the Middle Ages, required wine for its liturgy, abbots of monasteries and bishops throughout Europe also became wine growers, cultivating several vineyards. In France, Chateau Clos de Vougeot, Chateau Romanee, Chateau Chateauneuf-du-Pape, Johannisberger and many others.

Kakheti and the Alazani Valley is the most famous wine-making region in Georgia. This is also where Alaverdi, an active male monastery, is still located and home to the world's oldest still-functioning wine cellar.[6]

Beer was also brewed primarily in monasteries and, on a smaller scale, in individual households. Trappist beer is still brewed today by Trappist monks. Fourteen monasteries—six in Belgium, two in the Netherlands, and one in Austria, Italy, England, France, Spain and the United States.

The Eastern Orthodox and Roman Catholic churches and their calendars also influenced greater Europe and the diets of the Middle Ages. Most notably, meat consumption, as the intake of meat, was prohibited for almost one-third of the year. Fridays were fast days, and fasting was observed during both Lent and Advent. The only animal product allowed during these times was fish.

By the 16th century, Monasticism and its role in food and agricultural development had changed with the dissolution of the monasteries, which was introduced in 1536 CE by Henry VIII of England. This act led to the closure and confiscation of the lands and wealth of all monasteries in England and Wales. Monasteries with land ownership of over a quarter of all the cultivated land in England were sold off to families who supported Henry's break from the Catholic Church in Rome.[7]

In Europe, Martin Luther published "De votis monasticis" (On the Monastic Vows) in 1521, a treatise that declared monastic life had no scriptural basis. It was actively immoral, as it was not compatible with the true spirit of Christianity. For Luther, the Monasticism of the time directly opposed the "priesthood of all believers," implying, if not explicitly claiming, that there was no distinction in the righteousness of the peasant or priest, beggar or bishop.[8]

From the early sixteenth through to the eighteenth centuries, the percentage of land in church hands declined in Europe. The Protestant Reformation led to the seizure and sale of numerous former Catholic properties in the Holy Roman Empire, Scandinavia, the Baltic region, and the Low Countries. By the end of the English Reformation, only about 4 percent of the land remained in church hands, and almost all properties had been acquired by private buyers from the gentry or merchant classes.

Just as there was no single reason for humankind to adopt agriculture 10 000 years ago, no single cause can be attributed to the start of the Industrial Revolution in Britain in the late 18th century. Of course, food, in this case, the increase in agricultural output, played a significant role in its development.

The Second Agricultural Revolution, also known as the British Agricultural Revolution, occurred in England during the 17th and early 18th centuries. From there, it spread to Europe, North America, and the rest of the world.[9] It involved introducing new crop rotation techniques, such as incorporating turnips and clover over a four-year crop rotation sequence. Selective breeding of livestock also led to a marked increase in agricultural production.

Over this century, agricultural output in Britain grew faster than the population. After that, agricultural productivity remained among the highest in the world. The increase in food supply contributed to the rapid growth of the population from 5.5 million in 1700 to over 9 million by 1801. Domestic production gave way to food imports in the 19th century as the population more than tripled to over 32 million.[10]

The rise in productivity accelerated the decline of the agricultural share of the labour force, adding to the urban workforce on which a new era of industrialisation depended. This second, or UK Agricultural Revolution, is often cited as a foundation of the Industrial Revolution. As seen in the previous sections on grains and bread, the Industrial Revolution completely changed how we live and the industrial production of food.

In 1809, Nicolas Appert, a French chef known as the father of food preservation, developed the concept of preserving food through a heat process and hermetically sealing it in glass jars. He created this method in response to a request from the French government to find a way to preserve food for the military.[11] Appert's process involved sealing food in a jar or bottle, heating it to a specific temperature for a designated time, and then keeping the container sealed until it was ready for use. Although he may not have understood the scientific principles behind his method, which were explained by Louis Pasteur 50 years later, Appert won Napoleon's offered prize of 12 000 francs for improving food preservation techniques and enabling his army to be better supplied during military campaigns.

In 1813, Bryan Donkin, a Northumberland engineer, established the world's first canning factory in the United Kingdom, having acquired the patent from Peter Durand, who had registered the original patent in 1810 for using tin instead of glass.[12] The first tin cans were made of iron coated with tin to prevent rusting and sealed with lead solder. Today's food cans are made from steel or aluminium, depending on the type of can and the manufacturing method. Both steel and aluminium are non-toxic and recyclable.

Gradually, the production of canned foods became mechanised and the retort system, a room-sized steam autoclave used in canning, was invented in 1877. The retort was a key development in the canning process, enabling mass production and the safe preservation of food in cans through the application of pressure and high temperatures. Canned food is still popular today because it is incredibly convenient, affordable, has a long shelf life, and is readily available year-round.

REFERENCES

1. P. B. Newman, *Daily Life in the Middle Ages*, McFarland, 2018.
2. R. Wrangham, *Catching Fire: How Cooking Made Us Human*, Profile Books, London, 2009.
3. R. R. Davies and B. Smith, *Lords and lordship in the British Isles in the late Middle Ages*, Oxford University Press, New York, 2009.
4. C. H. Lawrence, *Medieval monasticism: forms of religious life in western Europe in the Middle Ages*, Longman, London, New York, 1984.
5. Aleteia, The Catholic monasteries that invented our favourite cheeses, https://aleteia.org/2022/04/06/the-catholic-monasteries-that-invented-our-favorite-cheeses.
6. G. M. Taber, *In Search of Bacchus*, Simon and Schuster, 2009.
7. M. Cartwright, *Dissolution of the Monasteries*, World History Encyclopedia, 2020, https://www.worldhistory.org/Dissolution_of_the_Monasteries/.
8. M. A. Lamport and M. E. Marty, *Encyclopedia of Martin Luther and the Reformation*, Rowman & Littlefield, Lanham, Maryland, 2017.

9. P. Brassley, M. Winter, M. Lobley and D. Harvey, *The Real Agricultural Revolution*, Boydell Press, 2023.
10. E. Kerridge, *The Agricultural Revolution (Economic History)*, Routledge, 2013.
11. N. Appert, *The Art of Preserving All Kinds of Animal and Vegetable Substances for Several Years*, 1811.
12. M. Greenland and R. Day, *Bryan Donkin*, 2016.

CHAPTER 7

A Cornucopia of New World Food

7.1 THE COLUMBIAN EXCHANGE

On August 3, 1492, Italian explorer Christopher Columbus, sponsored by the Catholic monarchs King Ferdinand II and Queen Isabella I of Spain, departed from Palos, Spain, and began his voyage across the Atlantic Ocean with a crew of just 90 men and three ships.[1]

His objectives were threefold: to find a shorter sea route to Asia—trade between Europe and Asia was highly lucrative, with the silk trade and Asian food spices; to make his fortune; and to spread the Catholic religion on behalf of his sponsor. In this regard, he failed to achieve his objectives on all counts. However, his expedition set in motion a widespread transfer of people, plants, animals, diseases, and cultures that would affect nearly every society on the planet.

"The Columbian Exchange", as American historian Alfred Crosby coined the term, was heavily weighted in Europe's favour with the successful appropriation of New World land and staple crops such as corn, potatoes, and cassava.[2] In return, the New World received some of the most devastating diseases that continuously plagued Europe, Africa, and Asia populations. The local population—never before exposed to vicious Old-World pathogens like smallpox, measles, mumps, whooping cough,

---

Food and Us: The incredible story of how food shapes humanity
By Seamus Higgins
© Seamus Higgins 2025
Published by the Royal Society of Chemistry, www.rsc.org

influenza, chicken pox, and typhus and thus lacking any immunity to them—began dying at apocalyptic rates.

Many historians now believe that new diseases introduced after Columbus's arrival killed off as much as 90 per cent of 40–100 million Indigenous people. The loss is considered among the most significant demographic disasters in human history.

By the 20th century, almost one-third of the world's food supply came from plants first cultivated in the Americas. Think of Vanilla and Avocado. Tomatoes, initially prized in Italy for their ornamental value, from the 19th century, tomato sauce became typical of Neapolitan cooking and, ultimately, Italian food in general. Introduced to India by the Portuguese, chilli and potatoes from South America have become integral to Indian cuisine.

Coffee, introduced in the Americas *circa* 1720, originated in Africa and the Middle East. Sugar cane (introduced from South Asia in the early 16th century) became the primary export commodity of extensive Latin American plantations.

Before the Columbian Exchange, there were no horses, cattle, sheep or pigs in America. There were no oranges in Florida, cattle in Argentina, bananas in Ecuador, paprika in Hungary, tomatoes in Italy, potatoes in Ireland, coffee in Colombia, or pineapples in Hawaii. There were also no rubber plants, maize, or cassava in Africa, chilli peppers in Thailand, and chocolate in Switzerland.

The global ramifications of the Columbian Exchange were not limited to Europe and the Americas. It also had a significant impact on Africa and, to a lesser extent, Asia. The discovery of quinine aided European exploration and eventual colonisation of the vast tropical regions of these continents. Extracted from the bark of the Cinchona tree, the indigenous population of the Andes used the bark to treat malaria and other types of fever.[3] In 1820, French scientists Pierre Pelletier and Joseph Caventou developed a process to extract quinine from Cinchona bark, significantly enhancing the medicine's potency. This discovery came just in time for European empires as they expanded into malaria-ridden parts of the world.

The cultivation of cash-rich crops in the Americas, along with the devastation of native populations from disease, resulted in a demand for labour that was met with the abduction and forced slave labour of over 12 million Africans during the 16th to 19th centuries.

The Columbian Exchange also vastly expanded the scope of production of some new and soon-to-be widespread global addictions, from tobacco to food, bringing the pleasures—and consequences—of coffee, sugar, and chocolate consumption to millions of people.

7.2 DEVELOPING A SWEET TOOTH

7.2.1 Sugar

As we know it today, raw sugar has been around for a long time; in 510 BCE, Emperor Darius of what was then Persia invaded India, where he found "the reed which gives honey without bees".[4]

Sugar was only later discovered by Western Europeans as a result of the Crusades in the 11th century CE. Crusaders returning home talked of how pleasant this "new spice" was. The Moors had also introduced sugar cane and other new crops, including oranges, lemons, and peaches, to Southern Spain. In 1319 CE, it was recorded that sugar was available in London at two shillings per pound. Today's prices would equate to about one hundred pounds per kilogram; as can be seen, it was a very expensive and luxurious "spice."[5]

In the mid-16th century, Bartolomeo Scappi was arguably the most famous chef of the Italian Renaissance.[6] He oversaw the preparation of meals for several Cardinals. He was considered a master of his profession, eventually becoming the personal chef of Pope Pius IV, a member of the Medici family. Scappi published Europe's first printed cookbook, De honesta voluptate et valetudine ("On honorable pleasure and health"), which included recipes for eating "healthy" food while enhancing its taste.[7]

Given who his patrons were, money was obviously no object, and his recipe book included many recipes for dishes in which sweetness was the predominant taste. In addition to seasoning meat, vegetables, and pasta dishes with as much as half a pound of sugar, he also detailed fruit pies and tarts; sweet custards and cream dishes; all sorts of biscotti and cakes, and pastry-like conceits that made use of candied fruits or marzipan. His advice was to use a lot of sugar. "Nothing given to us to eat is so flavourless that sugar does not season it," Scappi wrote. It would take another three centuries before the world followed his advice!

The supply potential of this "new spice" in the early 16th century encouraged Portuguese entrepreneurs to export enslaved people to newly discovered Brazil, where they rapidly started growing highly profitable sugar cane crops. Enslaved Africans were imported in such huge numbers that by the end of the 17th century, up to half of Brazil's population consisted of enslaved Africans.[8]

By the 1680s, the Dutch, English and French all had their own sugar plantations and slave colonies, with sugar production surpassing even that of Brazil. For a while, British-owned Barbados became the largest sugar producer in the world, only to be beaten by Jamaica and French-held Santo Domingo in present-day Haiti.

In the 18th century due to the increasing popularity of tea and coffee, both naturally bitter, sugar became popular as an indispensable sweetener. In the late 19th century, cheap jam (one-third fruit pulp to two-thirds sugar) began to appear on the table of every working-class household. The growing demand for sugar in Britain and Europe encouraged further growth and profit earnings, creating the term "white gold".[9]

The figures are astonishing. Britain's annual per capita sugar consumption was 4 pounds in 1704, 18 pounds in 1800, and 90 pounds in 1901—a 22-fold increase to the point where Britons had the highest sugar intake in Europe. And while slavery had been abolished (lastly, in Cuba, in 1884), cheapness was sustained by new flows of indentured labour from India, Africa, and China. Even the UK government took its cut by introducing the Sugar Act in 1764, effectively creating a 34 percent sugar tax on all imported sugar.[10] The same tax was abolished by Gladstone in 1874, and sugar gradually changed from a luxury item to an everyday commodity.

Britain's naval blockade of Napoleonic France at the start of the 19th century prodded the French to seek an alternative to Caribbean sugar supplies. It gave birth to the European sugar beet industry. The 20th century has seen this traditionally heavily subsidised and tariff-protected industry grow to produce approximately 50 percent of Europe's sugar, including the UK, which now consumes around 2.0 million tons of beet (60 percent) and cane sugar (40 percent) annually.[11]

7.2.2 Chocolate

When Cortes returned to Spain in 1528, from the Americas, unlike Columbus, Cortes brought the beans, the recipe, and the equipment necessary to make the Aztec's original chocolate beverage. It is suggested that the Spanish created the word Chocolati, instead of using the original Aztec word Cacahuati. The True History of Chocolate authors Sophie and Michael Coe surmise that the Spanish substituted the Aztec word because they were uncomfortable with a thick, dark brown drink that began with the Spanish word "Caca"![12]

In 1615, cocoa made its way into the court of King Louis XIII, and in 1657, the first English chocolate houses opened, much like today's coffeehouses. In 1829, a Dutch chemist invented the cocoa press, which enabled cocoa butter to be extracted from the bean, leaving the powder we now know as cocoa. In 1847, Joseph Fry of Bristol developed a method for combining cocoa powder, sugar, and cocoa fat into a paste that could be moulded into a bar. John Cadbury created a similar product in 1849.[13] By today's standards, the original bittersweet chocolate bars of both Fry and Cadbury would not be considered very palatable.

That only happened in 1875 when Daniel Peter, a close friend and neighbour of Henri Nestlé, of Vevay, Switzerland, succeeded in mixing cocoa paste with Nestlé's already developed sweetened condensed milk, creating the world's first milk chocolate bar. In most countries milk chocolate products are now more popular than plain chocolate.

In 1879, another Swiss chocolatier, Rudolf Lindt, inadvertently left his mixing machine on over the weekend and "discovered" a much smoother textured chocolate. He developed a new mixing machine resembling a conch shell to process his chocolate. "Conching" chocolate became a global standard for quality chocolate manufacture.[14]

The scale up of chocolate production and product nuances over the past 100 years has been astounding. In the 20s, Mars introduced the best-selling chocolate bar of all time, Snickers, featuring added nougat, caramel, and peanuts. Think of other novel chocolate bars like Aero: Are those bubbles random or sized to specification?

Whether it's the ratio of 1g of fat to 2g of sugars found in milk chocolate or the fact that it appeals to all five senses (sight, touch, hearing, smell, and taste) thanks to its attributes: colour, snap, mouthfeel, and complexity of flavours and aromas, there can be no denying that People all around the globe love it.

To this day, the Swiss remain the largest consumers of chocolate, averaging 8.8 kilograms per head per annum. Globally, consumption is growing at a 2.7 percent CAGR and the sales value is expected to reach approximately US$11.02 billion (with an approximate sugar content of 50 percent) by 2029.[15]

7.3 PRESERVES AND CONDIMENTS

Pectin in fruit was first isolated by French chemist Henri Braconnot in 1825. It was named after the Greek pektikos, meaning congealed or curdled.[16] It is a polysaccharide, similar to cellulose and starch, and comprises long chains of sugar molecules. In fruit, pectin is concentrated in the skin and core, acting as a structural "cement" in the plant cell walls. In a jam, pectin forms a mesh that traps the sugary liquid and cradles suspended pieces of fruit.

Branches that extend from the long chains of pectin bond with each other to form the three-dimensional network that jam makers desire. In solution, these branches are reluctant to bond because they attract water molecules, which stops them from bonding, and because they have a slight negative electrical charge, they repel one another. To solve the first problem, sugar is added, which binds to the water molecules and frees up the pectin chains to form their network. The negative charges are reduced by the naturally occurring acid in the fruit or by adding lemon juice or a similar substance to the mixture.

When sugar prices fell in the late 19th century, jams and marmalade (one-third fruit pulp to two-thirds sugar) became a staple of the masses. They were used to enliven the dark wholemeal bread eaten by the working classes.

At about the same time, Henry Heinz saw a market for ketchup in the United States. Ketchup had been an immensely popular addition to fish, meat, vegetables and gravies since the 18th century. It was a British staple imported to the United States and was well-established by the 19th century as a kitchen necessity. However, it lacked consistency and quality assurance. In addition, overcooking, spoilage, and large quantities of camouflaging spices often destroyed taste and appearance.[17]

Heinz utilised a reputation for delivering an unadulterated product of high and consistent standards—in a clear bottle—to seize the market. Heinz's ketchup was born from combining the pectin of ripe red tomatoes with a sugar and vinegar mix, setting a new benchmark for ketchup standards and taste profiles.[18] Today, Heinz is the market leader in ketchup, selling over 1.8 million bottles per day, which translates to around 650 million bottles sold globally each year.

REFERENCES

1. L. Bergreen, *Columbus: the four voyages, 1492–1504*, Viking, London, 2013.
2. A. W. Crosby, The Columbian exchange (by) Alfred W. Crosby, Jr, *Foreword by Otto von Mering: biological and cultural consequences of 1492*, Greenwood Pub. Co, Westport, Conn, 1972.
3. F. Rocco, *The miraculous fever tree: malaria and the quest for a cure that changed the world*, Harpercollins, New York, Ny, 2003.
4. A. A. Stern, *Raw Sugar*, Xlibris Corporation, 2020.
5. J. Braithwaite, *From cane to sugar*, Lernerclassroom, 2004.
6. V. von Hoffmann, *From Gluttony to Enlightenment*, University of Illinois Press, 2016.
7. L. Costa, Luxury and sloth, https://stravaganzastravaganza.blogspot.com/2013/05/luxury-and-sloth.html.
8. D. W. Tomich, R. F. Monzote, C. V. Fornias and R. de Bivar Marquese, *Reconstructing the Landscapes of Slavery*, UNC Press Books, 2021.
9. J. Walvin, *Sugar: The World Corrupted: From Slavery to Obesity*, Simon and Schuster, 2018.
10. S. Marcus, Sweetness and Power: The Place of Sugar in Modern History, Sidney W. Mintz, *Am. Ethnol.*, 1986, **13**(2), 377–379.

11. EU, Sugar Overview, https://agriculture.ec.europa.eu/farming/crop-productions-and-plant-based-products/sugar_en.
12. S. D. Coe, M. D. Coe, *The True History of Chocolate*, Thames & Hudson, 2013.
13. S. T. Beckett, *The science of chocolate*, Royal Society Of Chemistry, Croydon, 2019.
14. S. Higgins, 100 years? Piece of cake, https://www.thechemicalengineer.com/features/100-years-piece-of-cake/.
15. Statista, Chocolate Confectionery – Worldwide, https://www.statista.com/outlook/cmo/food/confectionery-snacks/confectionery/chocolate-confectionery/worldwide.
16. L. Wildsmith, *Handbook of Preserves*, The Crowood Press, 2023.
17. A. F. Smith, *Pure ketchup: a history of America's national condiment, with recipes*, University Of South Carolina Press, Columbia, 1996.
18. H. J. Heinz, *Ketchup Developer*, Checkerboard Library, An Imprint Of Abdo Publishing, Minneapolis, Minnesota, 2018.

CHAPTER 8

The 20th Century: Industrialised Agriculture

By the 20th century, the die had been cast for further change through the agricultural and industrial revolutions in Britain, America, and, to a lesser extent, in developing nations. The rapid evolution of machinery and subsequent farm mechanisation, combined with a better understanding of crop genetics, introduced dramatic changes in the intensity of agricultural food production and how it was farmed.

In tandem with these developments, the use and advances in new farm chemical fertilisers, pesticides, and fungicides were also widely adopted to improve crop yields. The 20th century, in turn, defied all Malthusian predictions of exponential population growth restricted by linear food production growth.[1] On the other hand, forces manifested by globalisation, such as market and trade liberalisation, capital flow, and the financialisation of industrialised farming, have created a powerful oligopoly of just a small number of global players.

8.1 FARMING

During the 68th annual meeting of the British Association for the Advancement of Science in 1898, Sir William Crookes, echoing

Malthus, delivered a presidential address expressing his concern that the world population would soon exceed the global food supply. "My chief subject is of interest to the whole world – to every race – to every human being ...... England and all civilised nations are in deadly peril of not having enough to eat.[2]"

He reasoned that with the supply of Chilean nitrate deposits (better known as Guano) being quickly depleted, he called on the world's chemists and chemical engineers to develop a nitrogen fixation process to tap into atmospheric nitrogen.

Although many scientists laid the groundwork, it was two German chemists at BASF, Fritz Haber and Carl Bosch, who first discovered how to turn atmospheric nitrogen into ammonia.[3] The process converts atmospheric nitrogen (N_2) to ammonia (NH_3) through a reaction with Hydrogen (H_2) using a metal catalyst under high temperatures and pressures. The use of synthetically created fertilisers became a significant stimulus in transforming the global food system, enabling larger-scale industrial agriculture and higher crop yields.

Tractors first emerged in the early 19th century, when steam engines on wheels were used to power mechanical farm machinery *via* a flexible belt. It was in 1893 that John Froelich established the Waterloo Gasoline Traction Engine Company, becoming the first company to manufacture and sell gasoline-powered farm tractors under the brand name "Waterloo Boy". In 1918, John Deere & Company, a farm Equipment Company based in Moline, Illinois, purchased the Waterloo Gasoline Engine Company for $ 2 100 000. In 1947, they launched their Model 55, the first self-propelled combine harvester, considered today the forerunner of today's combine harvesters.[4]

Early 20th-century agriculture was labour-intensive, taking place on many small, diversified farms in rural areas. In 1900, US farms employed nearly half the entire US workforce and 22 million work animals to produce an average of five different commodities. The same agricultural sector of the 21st century is now concentrated on a small number of large, specialised farms in rural areas where less than a fourth of the US population lives.[5] These highly productive and mechanised farms now employ less than a 1.5 percent share of US workers and use 5 million tractors instead of the horses and mules of earlier days.

Following the Great Depression of the late 1920s and early 1930s, the American government introduced significant financial intervention into the agricultural sector. It began as a means of ensuring food security but has since evolved into a farming system that still relies on subsidies, grants, and other forms of support.

In 2022, the US federal government provided farms with $15.6 billion in subsidies or direct farm program payments.[6] Globally, the UN estimates that support for producers in the agricultural sector amounts to approximately US$ 540 billion per year, accounting for 15 percent of total farm production value.[7]

8.2 CROPS

As we have seen since the birth of agriculture some 10 000 years ago, farmers have always selected seeds for desirable traits in various crops. These desirable traits included crop varieties (cultivars) with shortened growing seasons, increased resistance to diseases and pests, larger seeds and fruits, nutritional content, shelf life, and/or better adaptation to diverse ecological conditions. Indeed, one wonders if the Egyptians had access to today's plant technology when they were trying to cultivate new wheat varieties some 3000 years ago, what kind of popular grains would we be eating today?

Traditionally, new varieties in plant breeding were developed by selecting plants with desirable characteristics or combining qualities from two closely related plants through selective breeding. Pollen with the genes for a desired trait is transferred from plants of one crop variety to the flowers of another, which also possess other desirable characteristics. Eventually, the desired trait will appear in the new plant variety through careful selection of offspring.

Using selective breeding, Norman Borlaug, an American biologist, created a dwarf variety of wheat that put most of its energy into edible kernels rather than long, inedible stems.[8] The result: more grain per acre. Similar work at the International Rice Research Institute (IRRI) in the Philippines, led by Borlaug, dramatically improved the productivity of the grain that feeds nearly half the world—from the 1960s to the 1990s, rice and wheat yields in Asia doubled.

Even as the continent's population increased by 60 percent, grain prices fell; the average Asian consumed nearly a third more calories, and the poverty rate was cut in half.

When Borlaug won the Nobel Peace Prize in 1970, the citation read, "More than any other person of this age, he helped provide bread for a hungry world."

In the 1920s, it was discovered that heritable mutations could be induced in plants through irradiation or chemical treatments of seeds. Plant breeders learned that they could make mutations happen faster with a process called mutagenesis. Mutations are changes in a plant's genetic makeup. They can also occur naturally, resulting in new beneficial traits. Most natural mutations are random and result from a change within the plant's cells triggered by cold weather, temperature fluctuations or insect damage.

The expectations for improvements in crop varieties from Mutagenesis were prominent in the 1950s to 1960s, and indeed, many varieties were released. The mutant variety database collects information on plant varieties created through radiation breeding and other techniques. It logs over 3000 improved varieties, including fruits such as grapefruit, bananas, as well as grains like rice, wheat, and barley.[9]

Since the 1980s, interest among plant breeders in using mutation breeding has declined, likely due to expectations of new genetic modification technologies (GM traits) and difficulties in managing the load of accompanying harmful mutations in selected lines, which hindered the development of high-yielding varieties based on mutations.

The primary distinction between selective breeding and genetic engineering is that selective breeding does not alter the organism's genetic material. In contrast, genetic engineering changes the genetic material of the organism.

With advances in mechanisation, mechanical harvesting, and access to much larger distant export markets, it has also become increasingly profitable for crop farmers to specialise in just one or two grain crops. While monoculture farming had existed for several years—think of sugar cane—so-called monoculture grain farming meant more efficient planting, growing, and harvesting methods requiring less expensive equipment and fewer labourers with specialised knowledge of individual crops.

The advent of genetically engineered grain crops also encouraged the practice of monoculture farming on an industrial scale.

It was in 1994 when GM crops were first introduced in the USA with the "Flavr Savr" tomato, which had been genetically modified to slow its ripening process, delaying softening and rotting.[10] The signature technology of this approach—and the one that brought both success and controversy to Monsanto—was genetically modified, or GM, crops. Since their first release in the 1990s, they have been adopted by 28 countries and planted on 11 percent of the world's arable land, including half the cropland in the US. Today, about 90 percent of the corn, cotton, and soybeans grown in the US are genetically modified.

Americans have been consuming genetically modified (GM) products for nearly two decades. But in Europe and much of Africa, debates over the safety and environmental effects of GM crops have blocked their use. GMOs have also raised concerns about intellectual property. Farmers cannot save genetically modified (GM) seeds and must purchase new, patented seeds yearly from their supplier.

The debate about the safety of GMOs in terms of potential environmental and health risks is complex. Proponents argue that no studies have proven GM to be harmful and that such crops have prevented billions of dollars in losses in the US alone, while also benefiting the environment. Opponents argue that more studies are needed to convincingly prove their safety.

Before Monsanto became the face of industrial agriculture, it had already courted controversy as a chemical company. It was founded in 1901 to manufacture the synthetic sweetener saccharin, originally only produced in Germany.

Monsanto was one of a handful of companies that produced Agent Orange, the herbicide and defoliant chemical used by the US military as part of its herbicidal warfare program during the Vietnam War from 1961 to 1971.[11] It also sold and manufactured DDT, PCBs, the controversial dairy cow hormone, rBGH, and the Aspartame sweetener.

Roundup®, containing the active ingredient glyphosate, was developed and introduced by the Monsanto Company in 1974. The Roundup revolution took off in 1996 when Monsanto started selling genetically modified seeds that produced crops resistant to the herbicide's attack on weeds. "Roundup Ready" became

Monsanto's trademark for its patented line of crop seeds genetically engineered to be resistant to Roundup.

In 2018, Bayer bought Monsanto for US$62 billion to grow its business on both sides of the Atlantic and beyond. But even if Monsanto's brands no longer exist, its legacy remains.

In June 2020, Bayer agreed on a total payment of $10.1 billion to $10.9 billion to resolve current and address potential future Roundup™ litigation. The company also settled Dicamba drift litigation for a payment of up to $400 million and most of its PCB water litigation exposure for approximately $820 million.[12]

Between 1900 and 2011, the global population increased from 1.6 billion to 7 billion. Despite such explosive growth, the world's farmers produced enough calories in 2012 to feed the entire population, plus an additional 1.6 billion people.[13] Primary crop production increased by another 53 percent between 2000 and 2019, hitting a record high of 9.4 billion tonnes in 2019. Half of global primary crop production now comprises just four crops: sugar cane, maize, wheat, and rice.[14]

8.3 LIVESTOCK

As per Jared Diamond's book, "Guns, Germs and Steel", of a possible 148 wild candidate animals (approximately 110 lbs or 45 kg) available for domestication, some 10 000 years ago, we were only able to succeed with just 14 animals of which the primary 5 were sheep, cattle, goats, pigs and horses.[15]

Others include the two types of camels, alpacas, donkeys, reindeer, buffalo, and yak.

All of these species share three common social characteristics: they live in herds, occupy overlapping home ranges and maintain a well-developed dominance hierarchy. Such social animals lend themselves to herding since they tolerate each other and instinctively follow a dominant leader, imprinting humans as that leader. On the smaller animal side, consider dogs and chickens sharing similar characteristics.

And so, for most of the above agricultural period, the same domestic animals became an integral part of humankind's cultural heritage and food supply. Domestic animals were needed to be able to work the land. They became reliable sources of food products such as meat and meat products, dairy products (milk and cheese), poultry products (meat and eggs), and non-food products such as fibre (wool, mohair, cashmere, and leather).

The "industrialisation" of animal farming was about to change this philosophy for many as the various models of factory farming developed in the 20th century and have since been optimised to minimise costs and maximise profit.

Before the First World War, keeping poultry was never much more than a cottage industry, providing meat and eggs for the kitchen table. In 1923, a Southern Delaware housewife named Cecile Steele ordered 50 baby chicks for domestic egg production. By mistake, the local incubator sent her 500 chicks. According to family legend, instead of returning them, she temporarily stored the chicks in a repurposed piano box. She simultaneously asked a local lumberjack to build her a bigger shed.

Over the next 18 weeks, Cecile brought 387 of those 500 birds to a 2-pound weight and maturity. She then sold them as meat to city hotels and restaurants for 62 cents a pound, turning a nice little profit. The following year, she ordered 1000 chicks and in 1926, her order was 10 000.[16]

Today, 2700 chicken houses in Sussex County on the Delmarva Peninsula of Delaware produce 200 million broilers (chickens grown for meat only) annually. The hybrid broilers reared now reach a market weight of 4 pounds in seven to eight weeks, compared to the 2-pound weight achieved by Cecile in 1923 at 18 weeks.

In 2021, approximately 20.4 million metric tons of broiler meat were estimated to have been produced in the United States, making it the world's leading producer of chicken meat.

The chicken farming industry has since soared with globalisation. It has many different production levels, including feed mills, hatcheries, growing farms, and processing plants.[17]

In China, the broiler industry grew by 591 percent in the 20 years after 1985. Kentucky Fried Chicken is now one of the nation's biggest restaurant chains.[18]

Likewise, in terms of egg production, China produced the highest number of eggs in 2022, at just under 600 billion. In contrast, India, the world's second-largest producer, had approximately 120 billion eggs that year.[19]

Around the mid-sixties, the United States, the United Kingdom, and other industrialised nations commenced factory farming of beef, dairy cattle, and domestic pigs. In 1990, intensive animal farming accounted for just 30 percent of world meat production; by 2005, this had risen to 40 percent. Analyses in 2019 estimate that, globally, over 90 percent of all farmed animals now live on factory farms, while in the United States, that figure rises to 99 percent.

Global pork production has increased fourfold over the last 50 years and is forecast to continue growing. There have also been dramatic changes in the dairy industry; the US currently produces 60 percent more milk from 30 percent fewer cows than it did in 1967. This is because each cow now produces over two and a half times as much milk as it did 50 years ago. Milk prices in 1930 were $0.26 a gallon. In 2021, adjusted for inflation, the current milk price is roughly 37 percent cheaper in dollar terms.[20]

Despite the rising population growth of the 20th century, humans are now easily outnumbered by our farm animals. The combined total of chickens (19 billion), cows (1.5 billion), sheep (1 billion) and pigs (1 billion) living at any one time is three times higher than the number of people.[21]

On the one hand, factory farming has dramatically reduced the cost of producing meat and other animal products. For example, according to US government figures, the price of chicken at $2.38 today is roughly two-thirds the cost of what it would have been in 1935, taking average inflation into account. On the other hand, the price of the same chicken and/or other animal products does not include the broader environmental and social costs associated with factory farming, such as the higher levels of land and river pollution, compromised animal welfare, increased public health risks from antibiotic/antimicrobial resistance, growth hormones, and the possibility of new zoonotic disease.

Animal welfare is a significant concern of factory farming (also known as Confined Animal Feeding Operations, CAFOs), particularly from a consumer's perspective. Today, advocates from a broad range of backgrounds are increasingly calling for a shift away from harsh industrial practices to create a more just, equitable food system. As per the Humane League, the definitions of animal cruelty used by the CEOs of big meat companies will differ drastically from those used by grassroots animal advocates. While producers often claim to root out inhumane treatment of farmed animals wherever possible, many advocates believe that factory farms are inherently cruel.

The World Organization for Animal Health (WOAH), formerly the Office International des Epizooties (OIE), is an intergovernmental organisation that coordinates, supports, and promotes animal disease control. It is recognised as a reference organisation by the World Trade Organization (WTO) and had 178 member Countries in 2013.

Since the early 2000s, the OIE's mandate has been extended to include animal welfare, a complex issue that has been causing increasing concern among consumers and broader society. They have redefined animal welfare based on how an animal copes with the conditions in which it lives. *I.e.*, an animal is in a good state of welfare if scientific evidence indicates it is healthy, comfortable, well-nourished, safe, and able to express innate behaviour without suffering from unpleasant states such as pain, fear, and distress.[22]

The first intergovernmental standards on this matter were approved by the General Assembly of OIE Member Countries in 2005. They cover a wide range of issues across several key areas related to terrestrial and aquatic animals, including animal production systems, transportation, methods of slaughter, and animal experimentation.

Obtaining consensus and persuading 180 countries to commit to promoting these standards at the national level, regardless of their cultural, religious, or economic circumstances, is a significant step forward on this critical issue. Likewise, as reflected in more recent OIE efforts, animal welfare should be increasingly recognised as a regulatory function of veterinary authorities. Providing curricula in regulatory veterinary medicine should also include the enforcement of animal welfare statutes, as an emerging area of practice in veterinary medicine.

In a later chapter, we will examine environmental risks, concerns, and costs; for now, let's discuss the use of antimicrobials, including antibiotics, in factory farming. These products have become an integral component of factory farming. They are widely used for disease prevention and growth promotion in food animals.

In the United States, antimicrobial use in food animals is estimated to account for ~80 percent of the nation's annual antimicrobial consumption. A significant fraction of these products involves antimicrobials that are crucial in human medicine for treating common infections and are necessary for performing medical procedures, such as major surgeries, organ transplantation, and chemotherapy.[23]

Animal antimicrobial consumption is now nearly triple that of humans, and steroidal-based growth implants are widely used to stimulate feed intake and protein deposition to improve animal growth and feed utilisation. While meat production, since 2000, has reached a plateau in high-income countries, it has grown by 64, 53, and 66 percent in Asia, Africa, and South America, respectively. The global average annual animal consumption of antimicrobials per kilogram is estimated to be 45 $mg\,kg^{-1}$, 148 $mg\,kg^{-1}$, and 172 $mg\,kg^{-1}$ for cattle, chicken, and pigs, respectively.[24]

Based on current consumption, the same research team estimates that between 2010 and 2030, the global consumption of antimicrobials will increase by 67 percent, from $\pm 1\,560$ tons to ± 3605 tons. Consumption levels vary considerably between countries, ranging from 8 mg per PCU (a kilogram of animal product) in Norway to 318 mg per PCU in China. In both relative and absolute terms, China is the world's largest consumer of veterinary antimicrobials.

The problem with the overuse of antimicrobials is the creation of so-called superbugs, now commonly referred to as antimicrobial resistance (AMR). AMR occurs when bacteria, viruses, fungi and parasites change over time and no longer respond to medicines, making infections more challenging to treat and increasing the risk of disease spread, severe illness and death. While these new pathogens are often said to have recently emerged, developed, or learned to evade antimicrobial drugs, that is not the case. It would be more accurate to say they have evolved their own resistance to antimicrobials.[25]

Unlike humans or plants, bacteria do not have to wait for random mutation to develop an immunity to a particular antibiotic, such as penicillin. Suppose another species or strain has already gone through the process of mutation. In that case, the appropriate resistance genes are already present in bacterial populations, and what is referred to as horizontal transfer occurs, accelerating the evolution of penicillin resistance.

For example, in 2003, biologists discovered a single MRSA plasmid—a genetic structure in a cell that can replicate independently of its chromosomes—that not only conferred resistance to penicillin, streptomycin, and two other antibiotics but also to a disinfectant commonly found in wet wipes.[26]

According to a 2021 United Nations report, "AMR has now reached epidemic proportions but remains largely out of sight for most of the general population."

For example, the rate of resistance to ciprofloxacin, an antibiotic commonly used to treat urinary tract infections, varied from 8.4 percent to 92.9 percent for *Escherichia coli (E. coli)* and from 4.1 percent to 79.4 percent for Klebsiella pneumoniae in countries reporting to the Global Antimicrobial Resistance and Use Surveillance System (GLASS).[27]

The 2009 global Swine flu pandemic was caused by a virus that originated in farmed pigs and developed the ability to infect humans. The virus appeared to be a new strain of H1N1 resulting from a previous triple assortment of bird, swine, and human flu viruses, combined with a Eurasian pig flu virus, leading to the term "swine flu". It started in Mexico, just a few miles from a significant concentration of intensive pig farms and as per CDC estimates, it caused 60.8 million illnesses, 273 304 hospitalisations and 12 469 deaths in the US alone.

While initially considered a relatively minor influenza pandemic, the next global pandemic attributed to a pathogenesis crossover from animals proved much worse.

At the time of writing, the COVID-19, SARS-CoV-2 pandemic has so far infected more than 570 million people globally, with an estimated death toll of 6.38 million people. We are still counting the economic costs and the health conditions of people affected by what is now referred to as "long Covid".

While the origins of the COVID-19 virus are still debated, the most common consensus from various scientists is that the virus originated from an animal wet market in Wuhan, China. The same market included 47 381 individual animals from 38 species (including 31 "protected" species), all kept in dreadful conditions and teeming with all kinds of other infectious diseases, ready for slaughter on demand if not sold as pets.[28] As per Prof. David Macdonald of Oxford University, with these vast concentrations of diverse species under one roof, it is a matter of time before some other unwelcome disease might skip into the human population.

According to global health estimates by WHO (World Health Organization), 20 percent of all global deaths can now be attributed to infectious communicable diseases. Of these, 50 percent of the top 5 killers were zoonotic in nature. Over the last two decades, we have witnessed a significant increase in the number of pandemics, including SARS (2003), Nipah and Swine Flu (2009), Ebola (2014), and now COVID-19, all of which have originated from animal-human interactions within the food system.

John Ruskin was an English writer, philosopher, art critic, and polymath of the Victorian era. His "Common Law of Business Balance" resonates when we consider how our animal-sourced food system has evolved and developed in more recent times.[29]

"There is hardly anything in the world that someone cannot make a little worse and sell a little cheaper, and the people who consider price alone are that person's lawful prey. It's unwise to pay too much, but it's worse to pay too little. When you pay too much, you lose a little money—that is all. When you pay too little, you sometimes lose everything."

REFERENCES

1. T. Robertson, *The Malthusian Moment*, Rutgers University Press, 2012.
2. M. Ridley, *The rational optimist: how prosperity evolves*, Harper Perennial, New York, 2011.
3. V. Smil, *Enriching the earth: Fritz Haber, Carl Bosch, and the transformation of world food production*, Mit, Cambridge, Mass, 2004.
4. L. Klancher, *The Art of the John Deere Tractor*, Voyageur Press, MN, 2011.
5. Anon, Studies of Changing Techniques and Employment in Agriculture, Works Progress Administration, 1938.
6. Congress U. United States Government Policy and Supporting Positions, Independently Published, 2021.
7. United Nations, *A multi-billion-dollar opportunity – Repurposing agricultural support to transform food systems*, Food & Agriculture Organization, 2021.

8. L. F. Hesser, *The Man who Fed the World*, Leon Hesser, 2006.
9. N. W. Simmonds, *Principles of Crop Improvement*, Longman Science and Technology, 1979.
10. J. Fernandez-Cornejo, *First Decade of Genetically Engineered Crops in the United States*, DIANE Publishing, 2009.
11. B. J. Elmore, *Seed Money: Monsanto's Past and Our Food Future*, W. W. Norton & Company, 2021.
12. The Guardian Newspaper, Bayer agrees to a settlement of $10.9 billion over Monsanto's weedkiller Roundup, Wed 24th June 2020.
13. J. Eise and K. Forster, *How to feed the world*, Island Press, Washington, DC, 2018.
14. Food and Agriculture Organization Of The United Nations, *World Food And Agriculture - Statistical Yearbook 2020*, Food & Agriculture Organization, S.L., 2020.
15. J. Diamond, *Guns, Germs, and Steel: The Fates of Human Societies*, W. W. Norton & Company, 1997.
16. P. Martin, The 500: How Cecile Steele began a multi-billion dollar industry, Blogger on the Broadkill, December 2016.
17. Statista, Chicken meat production worldwide in 2022 and 2024, by country, https://www.statista.com/statistics/237597/leading-10-countries-worldwide-in-poultry-meat-production-in-2007/.
18. The week. The £3 chicken: have we gone too far in our search for cheap meat? https://theweek.com/arts-life/food-drink/955197/chicken-gone-too-far-search-cheap-meat.
19. Statista, Leading egg producing countries worldwide in 2022 (in number of eggs in billions), https://www.statista.com/statistics/263971/top-10-countries-worldwide-in-egg-production/.
20. J. M. MacDonald, J. Law and R. Mosheim, Consolidation in U.S., Dairy Farming, USDA, July 2020.
21. A. Thornton, *This is how many animals we eat each year*, World Economic Forum, 2019.
22. T. L. Maple and M. A. Bloomsmith, Introduction: The science and practice of optimal animal welfare, *Behav. Process.*, 2018, **156**, 1–5.
23. National Research Council, Board on Agriculture. Panel on Animal Health, Food Safety, and Public Health, *The Use of Drugs in Food Animals*, National Academies Press, 1999.

24. J. Dewulf and F. Van Immerseel, *Biosecurity in animal production and veterinary medicine*, Cabi, Wallingford, 2019.
25. T. M. Uddin, A. Chakraborty, B. M. Khusro, R. M. Zidan, S. Mitra, T. S. Emran, K. Dhama, H. Hossain Ripon, M. Gajdács, M. U. Khayam, M. D. Sahibzada, J. Hossain and N. Niranjan Koirala, Antibiotic resistance in microbes: History, mechanisms, therapeutic strategies and future prospects, *J. Infect. Public Health*, 2021, **14**(12), 1750–1766.
26. J. Lin, K. Nishino, M. C. Roberts, M. Tolmasky, R. I. Aminov and L. Zhang, *Mechanisms of antibiotic resistance*, Frontiers Media SA, 2015.
27. WHO, Antimicrobial resistance, https://www.who.int/docs/default-source/antimicrobial-resistance/amr-factsheet.pdf.
28. D. MacDonald, *The wet market sources of Covid-19: bats and pangolins have an alibi*, Oxford Science Blog, 2021.
29. J. Ruskin, *The Works of John Ruskin*, 1905.

CHAPTER 9

The 20th Century and the Business of Food Production

9.1 FOOD PRODUCTION AS "ONE OF THE MODERN ARTS"!

It could be said that one of the main changes to affect the 20th-century industrial manufacturing industry, including food production, saw its beginnings when Henry R Towne, himself a mechanical engineer and director of the Yale lock company, presented an address to the American Society of Mechanical Engineers, entitled "The Engineer as an Economist."

Towne argued that there were sound engineers and good people in business, but they were seldom the same. He asserted that "the management of works" had become a matter of great and far-reaching importance, justifying its classification as "one of the modern arts".[1]

In the decades to come, "management", as we know it today, would emerge and shape the world of work and industry. Masters of the material business world, from Frederick Winslow Taylor to Michael Porter, Tom Peters, and Michael Hammer, would all have a disproportionate effect on business management in general, including the food industry.

Before the Industrial Revolution, there was little to no management, meaning that anyone other than the owner of an enterprise handled tasks such as coordination, planning, control,

---

Food and Us: The incredible story of how food shapes humanity
By Seamus Higgins
© Seamus Higgins 2025
Published by the Royal Society of Chemistry, www.rsc.org

reward, and resource allocation. Beyond a limited few kinds of organisations, such as the church and the military, a small number of large trading, construction, and agricultural enterprises existed. Still, little of what we would recognise today as so-called managerial practice was evident.

Management, as a science, and Taylorism *et al.*, began in earnest in the early 20th century.[2] It focused on increasing productivity and efficiency through standardisation, division of labour, mass production and a centralised hierarchy.

Consider Henry Ford and his Model T automobile. In 1913, he introduced the world's first moving assembly line for mass-producing his vehicles. His innovative approach reduced the time to build a car from more than 12 hours to just 1 hour and 33 minutes. The move from batch production to a continuously moving assembly line enabled him to reduce the cost of his 1909 Model T from $850 to $260 by 1924. Between 1913 and 1927, his factories produced more than 15 million Model Ts.[3]

Ford's efforts drew on the scientific management approach promoted by the American engineer Frederick Winslow Taylor. Other industries, including the food and agricultural sectors, quickly took notice and soon adopted the mass production assembly line.

Thus began the nascent Food industry's conflict with business management theories rooted in economics. Despite the fact that unlike other manufacturing industries, the basic tenets of food supply, at its core, have always been about supplying a basic human need.

Hence, the ethical implications of our food system evolving as part of our larger social, economic and political systems.

When Frederick Winslow Taylor published his Principles of Scientific Management in 1919,[4] as Oliver Stone Dene summarised in 1923, he set off a century-long quest for the right balance between the "things of production" and the "humanity of production".

Alternatively, as some would have it, between the "numbers people" and the "people's people".

Similarly, post-WW2, the debate around a company's purpose and its management's role, was it as per Peter Drucker, as per his more than 30 Harvard Business Review (HBR) essays on managerial practice. There is only one valid definition of business purpose: to create a "customer" with a more inclusive view of "stakeholder capitalism," including stakeholders, employees, customers, and the public at large—or was it more about creating wealth for shareholders by way of "shareholder capitalism"?[1]

With the former broad-minded concept steadily chipped away by partisans of the latter, a drive to mass-produce food profitably was also spurred on by the rise of competitive retail market chains and a steadily growing urban consumer market, creating a new market for readily prepared convenience foods.

In the 1970s, management focus changed as we think of big business and the beginning of the modern era on Wall Street, where the trendiest idea, at the time, was following "Sun Tzu's" gung-ho, 2500-year-old military manual, "The Art of War".[5]

It was also in 1970 that the economist and Nobel laureate Milton Friedman published his now-famous essay in The New York Times Magazine, arguing that a business's sole social responsibility is to increase its profits.

Strategic management became a new buzzword for size, growth, and portfolio theory.

Tools like strategic planning (GE) and the growth share matrix (BCG) were used to formalise strategic business planning processes.[6] Although several food analogies and terminology were used, such as "milking," low growth, high share, cash "cows," liquidating/divesting, low share, low growth, "pets," or "dogs," the BCG share matrix model was built on the logic that market leadership results in maintaining a superior profit return.

As the business environment grew increasingly competitive and connected with a blooming management consultancy industry, competitive advantage became a priority for organisations in the 1980s. Tools such as Total Quality Management (TQM), Six Sigma, and Lean were utilised to measure processes and enhance productivity.

At the same time, forces driven by globalisation, such as market and trade liberalisation, capital flows, and urbanisation, enabled this growth to succeed and altered the nature of our food systems by increasing the diversity and affordability of food while also changing its quality and nutritional value.

The 20th century saw the rise of other new disciplines and service-linked industries, such as asset investment agencies, management consultancy and the advertising and marketing industry. The latter focusing on understanding what drives customer behaviour and anticipating their needs and wants.[7] Delving into deep-rooted psychological desires to create a "need" for new food products and targeting specific demographics that are more likely to have that "need."

Greater Taylorism, launched in the era of slide rules, had found the means to create ever more precise models of how manufacturing businesses should perform. Business management as a "new science" or "one of the modern arts"?

In 2019/20, the United States conferred about 1 million postgraduate degrees, an increase of 21 percent since 2009/10.

Twenty-three percent of master's degrees were conferred in business (197 400), 17 percent in education (147 000), and 16 percent in health professions and related programs (135 500). By comparison, less than 1 percent (7544) of master's degrees were awarded in agriculture and natural sciences.[8]

9.2 AS ADVERTISED: TASTE, PRICE AND CONVENIENCE (HFSS)[†]

In 1865, at the end of the American Civil War, John Pemberton, a lieutenant colonel of the Confederate army, returned to his original profession as a pharmacist and moved to Atlanta, Georgia, USA. Having received a nasty sabre wound to his chest during the last major battle of the Civil War, Pemberton became addicted to one of the most widely prescribed painkillers of the time, morphine.[9]

After much experimentation in his pharmacy, he formulated a drink he believed might be able to cure his morphine addiction. Known as "Pemberton's French Wine Coca", he claimed it could be used to treat anything from morphine addiction to depression or alcoholism, impotence, and other conditions.[10]

Pemberton was probably inspired by another imported and massively popular drink of the day, Vin Mariani. This Italian French "medicinal" wine was made from wine and cocaine and marketed by a Parisian chemist named Angelo Mariani.[11] Global devotees of the tipple included Alexander Dumas, Emile Zola,

---

[†] HFSS refers to food and drink products that are high in fat, salt and/or sugar.

actress Sarah Bernhardt, Pope Leo XIII, Presidents William McKinley and Ulysses S. Grant, and countless monarchs, including Queen Victoria of England.

As per Vin Mariani, Pemberton's early formulations combined alcohol and coca leaves (cocaine). However, he also included Kola nuts (containing caffeine) and Damiana, a mildly psychedelic shrub used as an aphrodisiac that grows wild in Mexico, Central America, and the Caribbean.

When Atlanta instituted prohibition laws in 1885, Pemberton had to change the formula of his drink so that it no longer contained alcohol. He replaced the alcohol content with approximately 15 percent sugar syrup and added citric acid to temper the excessive sweetness. He was now marketing his new product as an "intellectual beverage" and valuable "brain tonic" through local chemists' soda fountains.

His bookkeeper and partner, Frank Robinson, chose a new name for the drink based on its two main ingredients and penned it in the flowing script that later became the "Coca-Cola" trademark, which Candler eventually registered in 1893.

John Pemberton died in 1888, having sold his interests to Asa Candler. Candler, also a chemist and astute businessman, is said to have purchased Pemberton's shares for $2300.

While Pemberton is still known as the inventor of Coca-Cola, Candler turned both the product and the brand into "an icon of pleasure" (and profit!).

When Candler moved his advertising account to the D'Arcy agency in 1906, Coca-Cola was already being marketed as a "Delicious and Refreshing" and a "Revives and Sustains" drink.

By 1929, it had become "the pause that refreshes".

In 1931, Coca-Cola/and the D'Arcy agency reinvented Santa Claus for all by commissioning Haddon Sundblom to develop advertising images using Santa Claus; the same Coca-Cola images of Santa, dressed in a red tunic, as a warm, happy character with rosy cheeks, a white beard and twinkling eyes still enthrals children to this day.

By 1969, Coke had become "the real thing"; in 2009, opening a bottle of Coke was to "Open happiness"! Today, Coca-Cola is a global brand that sells more than 1.9 billion servings of Coke products daily. Its global advertising budget over the past seven years (2014-2021) averages $4 billion per annum.[12]

On the breakfast cereals front, Dr John Harvey Kellogg, of Battle Creek Sanatorium fame, set about devising food cures for what he believed were the common ills of the day, particularly, "one of life's deadliest vices, masturbation". A devout Seventh Day Adventist, he believed a diet centred on bland foods, such as cereal, would lead Americans away from sin, hence his invention of "clean foods" like cornflakes and granola.[13]

A former patient of the sanatorium, Charles W. Post, added sugar to the flake recipe in 1897 and introduced his first dry cereal, a crunchy blend of wheat and barley he called Grape Nuts. His first cornflake product, "Elijah's Manna," was introduced in 1904.

In 1906, William K. Kellogg, JH's younger brother, inspired by CW Post's success, launched his own company, the Battle Creek Toasted Corn Flake Company. To the chagrin of his elder brother, he added sugar, salt, and malt to his flake recipes.[14]

Meanwhile, John Harvey continued to manufacture and sell his original sanatorium products through his company, Kellogg's. He also sued his brother over who had the right to use their last name.

After years of fighting, during which WK's Corn Flakes became increasingly popular, William eventually won the right to use his surname for Kellogg's Corn Flakes in 1928.

From the outset, WK differentiated his cereal from Post's products by marketing it as the "only genuine Toasted Corn Flakes" and included his hallmark signature on every box and advertisement.

He also began spending heavily on advertising. In 1907, "Wink Day", his first major advertising campaign, targeted New York and advised housewives to "wink at your grocer and see what you get." What one got was a free sample box of W.K.'s cornflakes. Within a year, his New York market had increased 15-fold, and he'd sold 1 million cases of cereal.

An embryonic advertising and marketing industry had found a new niche with breakfast cereal products and producers looking at new national and international markets.[15]

From moms to children, when Kellogg's launched its toasted rice cereal, the use of cartoon characters, Snap, Crackle, and Pop (supposedly, the sound of the cereal when milk is added), became a hallmark of Kellogg's advertising when later animated for TV.

Post, for his part, changed the name of his "Elijah's Manna" with packaging showing the prophet Elijah receiving food from a raven. A design choice that didn't sit well with some Christians, to the inoffensive sounding "Post Toasties" and removed the biblical figure from the box.

He eventually collaborated with Walt Disney to feature Mickey Mouse as a Post mascot. It is said that Post paid a million dollars for the opportunity in the 1930s, during the height of the Great Depression.[16]

While Dr John Kellogg was adamant about keeping sugar out of his cornflakes, it was probably for the best that he wasn't around for the launch of Kellogg's "Sugar Frosted Flakes" in 1952, with the cartoon character Tony the Tiger as the face of the product.

Tony's well-known "they're G-r-r-reat" slogan became an instant hit with children. Although virtually fat-free, the breakfast cereal has a 38 percent sugar content. It was only in 1983 that the word "sugar" was quietly dropped from the brand name to become "Frosted Flakes".

The same advertising firm that invented Tony the Tiger, Leo Burnett, is also credited with creating various other popular advertising mascots, including the Marlboro Man for Marlboro cigarettes, the Pillsbury Doughboy, and the Jolly Green Giant.[17]

Toucan Sam, also from the Burnett stable, was launched in 1963 as the cereal mascot for Fruit Loops, or should that be "Froot" Loops?

When first launched in 1959, Froot Loops were initially called "Fruit Loops." However, a lawsuit claimed that the product was misleading because it presented itself as a legitimate fruit product when, in reality, it was mostly sugar and contained no actual fruit.[18]

According to the Harvard School of Public Health, Kellogg's Froot Loops contain 41 percent sugar by weight. Since then, at least two federal judges have upheld the use of the word "Froot" on the basis that it cannot reasonably be interpreted as suggesting the presence of natural fruit, not least because "froot" is not a real word and real fruit does not come in loops.

Toucan Sam, the Froot Loops mascot, was initially introduced as a blue Toucan, a type of bird with a large beak coloured red, orange, and yellow as per the loop colours first introduced in the breakfast cereal. The original voice actor for Toucan Sam was Mel Blanc, and the original creator of the cartoon version of Toucan Sam was Manuel R. Vega.[19]

According to the TV cartoon, Toucan Sam possessed "magical" powers. He could fly through the jungle, sniffing out his favourite Froot Loops due to the cereal's "fruity" aroma.

To me, this little guy had miraculous powers and could only exist in the land of make-believe cartoons and advertising! Moreover, as the artificial flavouring used in the US and the natural flavours used in the UK were added and mixed with the cereal's other ingredients before extruding the loops.

In 2014, Food Beast decided to test this fact for themselves. They found in a blind taste test that no one could identify any of the Froot Loop colours based on taste or smell.[20] Indeed, Colours have conditioned us to expect certain flavours pre-emptively. Regardless of what our tongue may report, as we have evolved, we still use our visual sense to compile a flavour profile.

To this day, producers continue to heavily promote breakfast cereals, with an advertising-to-sales ratio four to six times higher than that of most other food categories.[20] It remains a significant part of Kellogg's global operations, with an average daily advertising spend of around $ 2 million. However, the way that advertising spend is currently allocated may be about to change!

At the time of writing, Kellogg's UK is busy with another court action, this time against the UK government. To help tackle the rising crisis of obesity in the UK, the government is looking to limit in-store promotion of HFSS food products both in-store and online. From next year, it also intends to ban TV and online advertising of the same products before 9 PM.

Chris Silcock, MD of Kellogg's UK, said, "We believe the formula being used by the government to measure the nutritional value of breakfast cereals is incorrect and not legally implemented, as it measures cereals dry when they are almost always eaten with milk." In July 2022, the Royal Courts of Justice ruled against Kellogg in favour of the government. Post-judgment Chris Silcock still argued, "It makes little sense to us that consumers will be able to buy other products, like doughnuts and chocolate spreads, on promotion – but not many types of breakfast cereals."

Some processed foods predate the beginning of the 20th century, including canned soups, fruits, and vegetables, sweetened condensed milk, gelatin, ketchup, and other prepared condiments. It was after the First World War that these and other products found their way into the kitchens of eager young homemakers, with manufacturers promoting their innovative products through free recipe books, magazine advertisements, and newspaper advertisements.

In 1924, Clarence Birdseye, an American inventor and entrepreneur, founded Birdseye Seafood, Inc. to market his new "flash-frozen" technology foods, including fish, meat, and vegetables. It was voted the most meaningful innovation in humanity's food history by the Royal Society, the UK's National Academy of Science, based on ranking according to four criteria: accessibility, productivity, aesthetics, and health.[21]

Birdseye eventually sold his concept to General Foods, where he stayed as a consultant. Although the idea of frozen foods was not new, Birdseye's new method was the idea that a broad line of perishable foods—meats, poultry, seafood, fruits, and vegetables—could be dressed ready to cook, conveniently packaged, quick frozen, and delivered while still genuinely fresh.[22]

In 1954, the US Company Swanson put Birdseye's assertion to the test and paved the way for the modern-day ready meal when it introduced the frozen "TV dinner." Utilising the concept of a three-compartment aluminium tray filled with turkey and sweet gravy over cornbread dressing, frozen peas and sweet potatoes. The same tray could double as both a cooking and serving tray. The packaging was cleverly designed to look like mini TVs of the time, tuning knobs and all.

Their marketing campaign targeted "harried women" who worked outside the home—or just wanted a break from the daily grind of preparing family suppers. They sold 10 million units in their first launch year and an additional 25 million the following year.[23] The concept also forever changed how Americans eat, with far more people eating informally in front of the TV instead of gathering nightly at the dining room table.

In 1937, as another world war threatened, the timing was perfect for the arrival of a new processed, canned meat product called Spam. This is not to be confused with the modern usage of the word "spam," such as junk email or unsolicited messages sent in bulk by email, also known as "spamming." The rubbish email reference for spam probably originates from a Monty

Python sketch, where the name of the canned pork product, SPAM, is annoyingly repeated and unavoidable.

The company proudly boasts a Spam Museum situated in Austin, Minnesota, which details how the Spam name originated. The brother of a Hormel executive, Ken Digneau, came up with the name as a portmanteau word for "spiced ham" (even though neither ingredient is used in Spam!).

The new Spam product was introduced on July 5, 1937. The recipe uses pork shoulder (considered an undesirable by-product of pork butchery), water, salt, sugar, of course, and sodium nitrate (for colouring). It remained unchanged until 2009, when Hormel added potato starch to sop up the infamous gelatine layer that naturally forms when fatty meat is cooked.[24]

Spam was guaranteed to make a name for itself when the US government included it in war rations to be shipped overseas to Allied troops. It was economical, had a long shelf life, needed no refrigeration, and was ready to eat straight from the can. Despite letters sent to military newspapers at the time describing Spam as "meatloaf without basic training" or "ham that didn't pass its physical", the same gelatine layer, as mentioned earlier, was also used by soldiers to lubricate their guns and waterproof their boots!

Eighty years later, Spam and its signature blue can are still going strong. More than 8 billion cans of Spam have been sold worldwide. By the end of WWII—and with thousands of American GIs returning home who would refuse to eat it—Spam saw its role start to shift away from a convenient protein source to a "sometimes-food" side dish.

But while the core of America pushed Spam to the side of their plates, the same canned meat became a culinary sensation in much of the Asian Pacific and Hawaii. Asia's present-day fondness for Spam stems directly from World War II and subsequent conflicts, during which an entire generation grew up with Spam.

The tiny American territory of Guam, a curious combination of military base and island paradise for holidaying Japanese, is Spam's most successful market. The average Guamanian eats 16 tins a year, and even the local McDonald's sells Spam burgers.[25]

Speaking of McDonald's, it was also during the Second World War that two brothers, Maurice ("Mac") and Richard McDonald, opened the first McDonald's restaurant in 1940 in San Bernardino, California.

In 1948, they decided to revamp their business and produce large quantities of food at low prices. To achieve this, the brothers limited their menu to hamburgers, French fries, drinks, and pies. They named their new efficient format the Speedee Service System. This included a self-service counter that eliminated the need for waiting staff and servers. Customers received their food quickly because hamburgers were cooked ahead of time, wrapped, and warmed under heat lamps. These innovations allowed the brothers to charge just 15 cents for a basic hamburger, about half the price of competing restaurants.

In 1954, a salesman for a milkshake blender supplier named Ray Kroc was intrigued by the brothers' need for eight shake mixers and visited their restaurant to see how a small shop could sell so many milkshakes. Realising the potential of their restaurant concept, Kroc became a franchise agent for the brothers. In April 1955, Kroc launched McDonald's Systems, Inc., later known as McDonald's Corporation, in Des Plaines, Illinois.

In 1956, Kroc also founded the Franchise Realty Corporation, following the advice of Harry Sonnenborn, who later became the first president and CEO of McDonald's. This new company bought or leased the land on which the franchisee stores were built and charged the franchisee a rental or a share of sales above a fixed amount. The "Sonnenborn model" remains in use today, and real estate now accounts for 99 percent of the company's assets. Kroc also created a Hamburger University, a training facility in Chicago, to ensure consistent standards across the entire chain. Their employees and franchisees could graduate in "Hamburgerology" and major in French Fries.[26]

The "golden arches," designed according to the McDonald's logo, originally started out as two 25-foot-high yellow sheet metal tapered parabolas or single arches incorporated at either end of a McDonald's restaurant as part of their overall architectural building design in the mid-50s.

When Kroc looked to modernise the McDonald's image in the sixties, he hired the renowned marketing guru of the time, Louis Cheskin, as a design consultant. Cheskin, a clinical psychologist and marketing innovator, focused on catering to what consumers felt, desired, and needed, rather than trying to manipulate those ideas. His ground-breaking book "Colour for Profit" initiated a "scientific" approach to colour and design.[27] "We associate red with festivity, blue with distinction, purple with dignity, green with nature and yellow with sunshine".

Cheskin was also a subscriber to Freud's theories on how sexuality drives human behaviour and believed those impulses to be a valuable tool in marketing. As such, he convinced Kroc to retain the original yellow arches and combine them to form the M we know today. He also argued that the same symbol would have Freudian applications to the consumer's subconscious mind and would be a valuable asset in marketing McDonald's food, likening the logo to "Mother McDonald's breasts."[28] A helpful association if you're replacing homemade food or trying to attract children (and their subsequent parents!) to your restaurant.

Like most fast-food chains, McDonald's targeted many of its promotions at children. For example, its Happy Meals, which included a toy, or Ronald McDonald, a clown-like mascot introduced in 1963 and designed to appeal to young children or be available for in-store children's birthday party celebrations.

Kroc believed that advertising was an investment that would, in the end, yield a significant return, and advertising has always played a vital role in the development of the McDonald's Corporation. It advertises extensively and regularly. In addition to using television, billboards, and signage, it co-sponsors major sporting events, ranging from the Olympic Games to World Cup soccer.

The iconic golden arch logo, according to a global survey by Sponsorship Research International, found that 88 percent could identify the arches. By comparison, only 54 per cent recognised the Christian cross.[29]

Whether it was the Freudian association with the logo, extensive advertising and marketing spend currently $1.25 m per day or increasing portions of cheap fast food (average portion sizes have more than doubled since McDonald's started in the 50s), the McDonald's chain is now the largest restaurant company in the world. They support more than 40 000 outlets operating in over 120 countries worldwide.[30]

Globally, the quick-service restaurant (QSR) or fast food market is estimated at just under $1 trillion per year and is growing at a compound annual growth rate (CAGR) of 6.05 percent. McDonald's is the market leader, holding a share of approximately 21 percent. Followed by Starbucks, KFC, Subway, and Domino's Pizza, respectively.[31]

After WW2, women (whether still employed or at home) were encouraged to embrace frozen, dehydrated, canned, readymade, and pre-prepared foods, which promised to save time in the progressive modern era and allow more time for new leisure options—including watching television. Of course, food manufacturers capitalised on this new medium, contracting with celebrities and even children's cartoon characters to promote an ever-growing number of new products on sponsored television shows, radio, and popular magazines.

Throughout the 20th century, the food industry worked to provide not only convenience but also ostensibly "wholesome" and cheaper substitutes for natural foods. The invention of hydrogenation was a chemical process that turned vegetable oils into solid fats. This product, coupled with the depression of the 1930s, which led to a shortage of animal fats, created the perfect catalyst for the Margarine industry to grow as a cheap alternative to butter.

High-fructose corn syrup (HFCS) is a liquid sweetener alternative to sucrose (table sugar) used in many foods and beverages. Early developmental work was conducted in Japan during the 1950s and 1960s, and the first commercial HFCS product was shipped to the food industry in the late 1960s.

While there is no denying that flavour, texture, and nutrients suffered with the advent of advertised convenience in the 20th century, people began to rely on these conveniences and changed their buying and eating habits to accommodate them. From a food industry perspective, significant changes were still to come.

9.3 FROM A NATION OF SHOPKEEPERS TO "EVERY LITTLE HELPS"

A quote often attributed to Napoleon was first written by Adam Smith, of "Wealth of Nations" fame, in 1776, when he discussed the merits/demerits of "Britain being a nation of shopkeepers".[32]

From the 1800s to the 1950s, independent retailers were a cornerstone of British life. It was commonplace for people to buy groceries daily and visit the corner shop as a routine part of life. Shop assistants served individual customers directly, and goods were weighed out by hand, cut up, sliced, bundled, and wrapped. Everything was tallied on a receipt or jotted on the front of the plain paper bag used for wrapping the goods. In the early 1930s, 90 percent of over 500 000 grocery shops in Britain were independently owned.[33]

When Clarence Saunders opened his first "Piggly Wiggly" store (a name chosen to intrigue people) on Sept. 11, 1916, in Memphis, USA, it was remarkable as being the world's first self-service store.[34]

The store proprietor promised to "slay the demon of high prices". Similar to the way Jack Cohen, founder of Tesco, acquired the nickname 'Slasher Jack' due to his ruthless cutting of retail grocery prices in Britain from the 1920s onwards.

Cohen's early business slogan was "pile it high and sell it cheap", to be replaced in the 30s by "YCDBSOYA" – You can't do business sitting on your arse. Having grown his food stores to over a hundred before 1939, post WW2 and the end of food rationing, he opened his first self-service store and supermarket in Britain in 1958.[35]

In 1964, following intensive lobbying by Jack Cohen, Ted Heath, a former Prime Minister of the United Kingdom, led a parliamentary motion to replace the existing Retail Price Maintenance Act with the Resale Prices Act. The former enabled suppliers to set minimum and maximum selling margins, while the latter considered all resale price agreements to be against public interest.

By 1969, there were 3400 supermarkets in the UK, of which 800 belonged to Tesco. By 2000, the market share of independent stores in the UK had dropped from 80 percent in 1900 to just 18 percent. While lower prices became the key driver for supermarkets to attract shoppers and gain market share, from a supplier's perspective, one now had the situation whereby just four major players were controlling 76 percent of the nation's grocery shopping for 26 million households.[36]

Anybody who has run the gauntlet of trying to list a new food product or agree to a price increase with a major supermarket chain knows the drill. It is common practice on a global basis! First, there is the 15–20 minute wait at the reception after appointment time before negotiating the new product line listing fees, and woe betides if the product has variants. Trade discounts, advertising assistance, Gondola ends (aisle-end display costs per store, per week), merchandising and demonstration assistance, and/or allowances. What about free stock for promotion? And so on.

"Oh, so you can't deliver nationally on a just-in-time basis to all our stores? Not a problem; we can arrange the distribution for you through our distribution centres (DCs) at a 20–30 percent margin."

"A price increase? Well, that could be planned for in six months' time," and if you insisted otherwise, your product was always at risk of being relegated to the bottom display shelf.

Supermarkets characteristically say that they merely respond to consumers' wants, and if they get it wrong, consumers will go elsewhere. However, they play a crucial role in shaping consumer demand. Because of their power as gatekeepers rather than passive transmitters of consumers' wishes, their gatekeeping role can work to the detriment of both consumers and suppliers alike.

From a consumer's perspective, the overall layout design, from the time you enter the store, is known as a planogram, more simply defined as a schematic drawing or plan for displaying merchandise to maximise sales.

The specific foot traffic flow patterns are always geared to the left, leaving your right hand free (90 percent of the time, allowing for lefties!) for loading more items. If you're popping into the supermarket for one or two items, such as milk, eggs, or bread, chances are they're located at the back of the store, with long aisles of further temptation to navigate before you find them. Complementary products are often packaged together as a single product, such as meal deals, at a discounted price to encourage sales.

Nothing is left to chance in the store; even the music mix, which includes oldies and more modern happy tunes, is strategically chosen to keep customers relaxed and in a buying mood.

Studies suggest that as much as 50 percent of all groceries are sold due to impulse purchases – and over 87 percent of shoppers make at least one impulse purchase. "Buy one, get one free" offers, discounts, and in-store promotional displays all play a crucial role in increasing your intended supermarket shopping. If you have entered the store hungry, forget about it! Returning to your primal instincts, you have likely already overfilled your cart with high-calorie food you wish you were eating right now.[37]

The UK supermarket sector was the subject of a comprehensive investigation by the Competition Commission during 1999–2000. In its report, the commission identified several practices which, when carried on by any of the major supermarkets, adversely affected the competitiveness of some of their suppliers with the result that the suppliers were likely to invest less and spend less on new product development and innovation, leading to lower quality and less consumer choice. The commission considered that this would likely result in fewer new entrants to the supplier market than otherwise.[36]

On a more positive note, the growth of the supermarket industry did assist in expediting a number of new technologies that have become commonplace in the food industry today, such as the barcode. According to GS1 UK, the British branch of the global barcode regulator, 100 stores were scanning at the till in 1984. By 1995, the figure was greater than 20 000.

In 1984, Tesco facilitated the world's first online grocery shopping experience when a UK pensioner selected three home-delivery items using a computer technology called Videotex. Using her TV remote and her TV linked to her telephone line, she chose from a drop-down menu of 1000 items and transmitted her live order to the local Tesco store.

In 1993, Tesco introduced its new slogan, "Every little helps."[38] In 1994, it teamed up with Dunnhumby, a husband-and-wife team specialising in forensic retail analysis, to consider launching a new customer loyalty card. After a three-month trial in twelve stores and a board presentation, the chairperson of the time, Lord MacLaurin, commented that what impressed him most about this new technology was that they knew more about his customers after three months than he knew after 30 years.[35]

Tesco's Clubcard was launched in February 1995, and its nationwide rollout was the foundation of Tesco's rise to become the dominant retailer in the UK and one of the largest in the world. With a 1 percent savings for customers and the Dunnhumby database behind it, the loyalty card provided Tesco with unprecedented insights into who its shoppers were and how they shopped. It was also one of the most important retail innovations of the 20th century.

Globally, as per Kantar and Deloitte the FMCG (Fast Moving Consumer Goods) category, comprising 141 store chains, makes up approximately 66.4 percent of total retail sales. Retail revenue was roughly $28.2 trillion in 2021 and is projected to reach around $32.8 trillion by 2026.[39]

As in many other countries, grocery retailing in the UK has faced significant challenges since the COVID-19 pandemic began in 2020. Demand for groceries skyrocketed as pandemic-related lockdowns were implemented, and dining out became impossible. Growth in online grocery shopping became turbo-charged. Amazon achieved the highest retail revenue growth among the world's top ten retailers, with year-on-year growth soaring by 34.8 percent.

According to Evan Sheehan of Deloitte, retailers who can deliver to consumers what they want, where they want it, and when they want it will continue to win.

REFERENCES

1. W. Kiechel, Harvard Business Review (2012, November), The Management Century, PMID: 23155998.
2. F. W. Taylor, *The principles of scientific management*, Harper & Brothers, 1919.
3. M. O'Hearn, *Henry Ford and the Model T*, Capstone, 2007.
4. F. W. Taylor, *The principles of scientific management*, Harper & Brothers, 1919.
5. D. Liu and A. Snyder, *The way of the Wall Street warrior: conquer the corporate game using tips, tricks, and smartcuts*, John Wiley & Sons, Inc., Hoboken, New Jersey, 2022.
6. The BCG Growth-Share Matrix: Theory and Applications: The Key to Portfolio Management, 50Minutes.com, 2015.
7. D. Anselmo, *Marketing Demystified*, McGraw Hill Professional, 2010.

8. National Centre for Educational Statistics, https://nces.ed.gov/programs/coe/indicator/ctb/graduate-degree-fields.
9. G. Pollock, *The Sad True story of John Pemberton and the invention of Coca-Cola Amazon*, 2022.
10. M. Pendergrast, *For God, Country and Coca-Cola: the Definitive History of the Great American Soft Drink and the Company That Makes It*, Basic Books, New York, 2000.
11. Company M, Coca erythroxylon (Vin Mariani.) Its uses in the treatment of disease, 1886.
12. A. Ciafone, *Counter-cola: a multinational history of the global corporation*, University of California Press, Oakland, California, 2019.
13. B. C. Wilson, *Dr John Harvey Kellogg and the Religion of Biologic Living*, Indiana University Press, 2014.
14. S. Machlin, *American Food by the Decades*, ABC-CLIO, 2011.
15. J. Mcdonough, *The Advertising Age Encyclopedia of Advertising*, Routledge, New York, 2002.
16. S. Hulett, B. Mclain and J. Musker, *Mouse in transition: an insider's look at Disney feature animation*, Theme Park Press, San Bernardino, CA, 2014.
17. N. Meyerson, *How Tony the Tiger, Pillsbury Doughboy and Michelin Man became cultural icons*, CNN Business, 2022.
18. B. York, Froot Loops Cereal (History, FAQ & Commercials), https://www.snackhistory.com/froot-loops-cereal/.
19. T. Lawson and A. Persons, *The Magic Behind the Voices*, University Press of Mississippi, 2009.
20. K. Frederickson and N. Vijayan, The Untold Truth Of Froot Loops, https://www.mashed.com/200000/the-untold-truth-of-froot-loops/; J. Berning and A. N. Rabinowitz, Targeted advertising in the breakfast cereal market, *J. Agric. Appl. Econ.*, 2017, **49**(3), 382–399.
21. The Royal Society, Royal Society names refrigeration most significant invention in the history of food and drink, https://royalsociety.org/news/2012/top-20-food-innovations/.
22. M. Kurlansky, *Birdseye: the adventures of a curious man*, Anchor Books, New York, 2013.
23. History, TV Dinner's Disputed Origins, https://www.history.com/news/tv-dinner-history-inventor.
24. D. Mundorf, The 9 Most Popular Myths About Spam, https://www.foodie.com/1629822/9-most-popular-myths-about-spam/.

25. O. Thring, Consider spam, The Guardian Newspaper, Tue 28 Sep 2010.
26. R. Kroc and R. Anderson, *Grinding it out: the making of McDonald's*, St. Martin's Griffin, New York, 2016.
27. L. Cheskin, *Color for Profit*, Ig, New York, 2016.
28. E. Schlosser, *Fast Food Nation: the Dark Side of the All-American Meal*, Mariner Books/Houghton Mifflin Harcourt, Boston, 2001.
29. S. W. Londoner, Bigger than Jesus? McDonald's Golden Arches more recognisable than Christian Cross 27th October 2014.
30. Statista Number of McDonald's restaurants worldwide from 2005 to 2023, https://www.statista.com/statistics/219454/mcdonalds-restaurants-worldwide/.
31. Fortune business Insights, Quick Service Restaurants Market Size, Source: https://www.fortunebusinessinsights.com/quick-service-restaurants-market-103236.
32. A. Smith, *The wealth of nations*, W. Strahan and T. Cadell, London, 1776.
33. A. Seth and G. Randall, *The grocers: the rise and rise of the supermarket chains*, Kogan Page, Londres, 2001.
34. M. Freeman, *Clarence Saunders & the Founding of Piggly Wiggly*, The History Press, 2011.
35. S. Ryle, *Making of Tesco: a story of British shopping*, Transworld Publishers Ltd, 2013.
36. Uk Competition Commission, The supply of groceries in the UK market investigation, 2008.
37. C. Jannson-Boyd, How shops use psychology to influence your buying decisions, The conversation, 2022.
38. L. Salem, *The Handbook of Slogans*, Crimson, 2012.
39. Deloitte, *Global powers of retailing*, Deloitte, 2022.

CHAPTER 10

20th-century Food Advice

10.1 REGULATING THE INDUSTRY

At the beginning of the 20th century, only 14 percent of the world's population lived in urban areas, and only 12 cities had a population of 1 million or more. By 1950, 30 percent of the world's population resided in urban centres, and the number of cities with over 1 million people had grown to 83.[1]

Food safety had become a significant concern at the turn of the century. Think of Lower East Side Manhattan, New York, where some 2.3 million people (two-thirds of the city's population) lived in tenement housing serviced by 25 000 open food carts.[2]

Food products, such as raw milk, improperly handled or unpasteurised, can carry various pathogens, including tuberculosis, brucellosis, diphtheria, scarlet fever, Q fever, and several gastrointestinal infections, including listeria, salmonella, and *E. coli*, among others. Prepared foods of the time also included items such as formaldehyde, coal tar, and copper sulfate, as well as other toxins used to mask spoiled food.

Britain revised its Adulteration of Food and Drugs Act in 1872. It was not until the turn of the century in the US that Dr Harvey Wiley nicknamed the "Father of the Pure Food and Drugs Act," and a team of scientists from the Agriculture Department's

Bureau of Chemistry set out to test food additives for safety and/or determine safe consumption levels.[3] It was not until 1910 that the mandatory pasteurisation of milk supplied to New York City became law.

In Britain, the first Act to legislate on the production and sale of milk, the "Milk and Dairies Act of 1914," was finally implemented in 1925. This Act provided the framework from which all food safety legislation in the UK is written; consider the term "food which endangers health" and it remains the primary domestic food law source today.[4]

Additional pieces of legislation relating to labelling were also passed. For example, the Food, Drug, and Cosmetic Act of 1938 introduced penalties for misleading or false advertising in the United States. It contained new provisions providing safer tolerances for unavoidable poisonous substances and authorising factory inspections. Food laws, like other cultural change elements in society, evolved one piece at a time in response to situations that caught the attention of society, government leaders, or legislators tasked with improving or policing the same broad food safety principles.

Various food producers' growing power and influence driving their own economic interests also became part of that mix! The so-called "Great Butter *vs.* Margarine War", which began in the late 19th century and was not settled until after World War II, is a prime example.[5]

In 1869, a French chemist patented a lower-priced bread spread made from beef fat called Oleomargarine. Realising the patent's commercial potential, he sold it to a Dutch butter company that later became part of the giant food manufacturer Unilever. The company remained in the margarine business for nearly 150 years until it sold its margarine brands for $8 billion in 2017.

Margarine arrived in the United States in the 1870s. To the universal horror of the American dairy industry, butter producers quickly responded to the perceived threat to their market.

According to a New York Times article of the time,[6] the dairy industry launched a marketing campaign to convince politicians and the public that margarine was unhealthy and was being improperly sold as butter. After passionate lobbying from the dairy industry, the federal Margarine Act was introduced in 1886.

It imposed a restrictive tax on margarine and required manufacturers to pay prohibitive licensing fees. By the early 1900s, Congress had also imposed a 10-cent-per-pound tax on all coloured margarine, and more than 30 states had banned coloration outright.

At the time, butter, made from the butterfat of cows' milk, was traditionally yellow, as derived naturally from the carotene contained in grass-fed cows. In contrast, in its natural state, margarine, without added colouring, was an off-white, greyish colour, similar to beef fat. In response to legislation, margarine companies began distributing them with a separate packet of yellow food colouring.

Shortages in beef fat supply, combined with advances in the hydrogenation of plant materials, later became a topical subject in their own right due to their high trans-fat content and soon accelerated the production of margarine from a combination of animal fats and hardened vegetable oils. The depression of the 1930s and the rationing of dairy products during World War II contributed to the increasing popularity of margarine. Finally, in 1950, Congress repealed the 1902 Margarine Act. From a food legislation perspective, the above saga begs the question of whether these laws are based on concerns for consumers' safety or the dairy industry's economic well-being.

On a more positive note, with growing scientific awareness of the benefits of iodine supplementation in reducing goitre, the first modern legislated food fortification program began in the 1920s, when Switzerland introduced salt fortification with iodine to prevent the consequences of iodine deficiency and subsequent thyroid problems.[7] This type of legislated food enhancement developed further in the 1930s and 1940s when specific deficiency disease syndromes were finally identified after the new grain milling technology was introduced at the turn of the century.

In 1940, the US Committee on Food and Nutrition (FDA) recommended adding thiamine, niacin, riboflavin, and iron to flour. The FDA also established a standard for enriched flour, which identifies specific nutrients and their required amounts for addition to any flour labelled as "enriched" to improve its nutritional value.[8]

White flour was first fortified with calcium in the UK in 1941.[9] The government introduced this to prevent rickets, which was common in women joining the Land Army.[10] Since the 1940s, white flour has also been mandatory fortified with iron, vitamins B1—thiamine and B3—niacin B1, B2, and margarine was fortified with vitamins A and D. Of course, wholemeal flour is exempt as the wheat bran and wheat germ included in the flour are natural sources of vitamins and minerals from the grain.

Following the Second World War, the Food and Agriculture Organization (FAO) and the World Health Organization (WHO) teamed up to construct an International Codex Alimentarius (or "food code"), which emerged in 1963.[11]

In the wake of various food scares related to salmonella and BSE (bovine spongiform encephalopathy or mad cow disease), Britain passed the Food Safety Act of 1990, later foreshadowed by the EU, legislating that any supplier of a branded product is responsible for the safety of that product. Likewise, all fresh produce sold in an unpackaged form is considered to bear the retailer's brand. Accordingly, reputation and financial resources are at stake if firms fail to prove due diligence in detecting and preventing problems in the food chain.

In response to the legal requirement of due diligence, the first food-safety-specific private standards were created in the early 1990s.[12] Following the creation of the ISO 9000 quality standard, the International Organization for Standardisation (ISO) developed ISO 22000, first published in 2005, which outlines the requirements for a food safety management system that can be certified.

Similarly, the first iteration of HACCP (Hazard Analysis and Critical Control Points) principles was developed by NASA (the US National Aeronautics and Space Administration) in the early 1960s, when it was investigating various ways to create safe food for space missions. The Codex Alimentarius Commission adopted it in 1993.

Throughout the globalisation of food safety, many developed countries have implemented their own robust regulatory systems for food safety. Although the primary goal of these systems has been to protect consumers from consuming food that could be harmful to their health, several broader objectives have also emerged in these countries over the last few years.

An expanding food safety view focuses on managing hazard risk throughout the farm-to-fork continuum, implementing suitable practices and traceability, addressing food terrorism and intentional adulteration, vulnerability to food fraud, and antibiotic resistance. Australia, New Zealand, Canada, the United States, and the European Union have recently refined their food safety practices.

In 2021, the UK updated its fortification legislation to include folic acid in its mandatory wheat flour fortification standards. At least 68 countries currently have mandatory wheat flour fortification with folic acid. Fortifying with folic acid improves folate levels in women. It reduces their children's risk of having brain and spinal defects called neural tube defects (NTDs).

10.2 CONTROLLING CHD[†], SATURATED FAT OR SUGAR?

In 1955, during the ongoing butter-margarine debate, the US state of Wisconsin arranged a "senatorial" taste test in which blindfolded senators were challenged to tell the difference between butter and margarine. Most could tell the difference, except for the vociferously pro-butter Senator Gordon Roselip, who insisted that the margarine he had tasted for the test was butter.[13]

It turned out later that Roselip's wife, worried about her husband's heart, had been sneakily substituting yellow margarine for butter for years at the Senator's dinner table.

Mr Roselip's wife, although perhaps uninformed over the natural causes of possible CHD, was putting into practice what she believed to be a preventative precautionary measure to ward off the chances of her husband suffering from a heart attack.

Her concern, similar to many other mid-century Americans, was that heart disease, an uncommon cause of death in the US at the beginning of the 20th century, had now become the most common cause, accounting for 1 in 2 deaths in America.

By the mid-century, improvements in food safety and sanitation and the development of vaccines and antibiotics had dramatically decreased infectious disease mortality, and concomitant life expectancy grew to 69.7 years.

However, heart disease, cancer, and stroke have now replaced contagious diseases as the leading causes of death. Indeed,

---

[†] Coronary heart disease.

prevention and treatment were so poorly understood that most Americans and physicians of that era viewed atherosclerosis, where arteries become clogged with fatty substances called plaques, as a "natural" and inevitable feature of ageing. Many accepted early death from heart disease as unavoidable.

With hindsight, it is not hard to see why rates of heart disease were climbing. Smoking levels were much higher than today, with roughly 45 percent of the American population self-proclaimed smokers. In the UK, there was four times as much tobacco smoked in the early 1960s compared with today.[14] Similarly, adopting a more sedentary lifestyle, characterised by the use of cars and TV, and consuming processed foods high in fat, salt, and/or sugar (HFSS) became the norm, leading to increased serum cholesterol levels.

In the early 1950s, University of California researcher John Goffman and his associates identified today's two well-known types of cholesterol: low-density lipoprotein (LDL) and high-density lipoprotein (HDL). When they studied plasma from postmortem heart attack patients, they also found a correlation between higher cholesterol-carrying LDL and the prevalence of coronary atherosclerosis with resultant coronary heart disease.[15]

HDL has since become known as the "good" cholesterol as it carries cholesterol to the liver, where it can be removed from the bloodstream before it builds up in the arteries. Conversely, LDL is considered the "bad" cholesterol, which leads to a buildup of cholesterol in the arteries.

In January 1961, an American epidemiologist named Ancel Keys made the cover of Time Magazine and was duly hailed as the man most firmly at grips with the cause of chronic heart disease (CHD). The inventor of the US wartime K (for Keys) ration pack proclaimed in the article that Americans eat too much fat.[16] "With meat, milk, butter and ice cream, the calorie-heavy U.S. diet is 40 percent fat, and most of that is saturated fat—the insidious kind, says Dr Keys, that increases blood cholesterol, damages arteries, and leads to coronary disease". "People should know the facts," he said. "Then, if they want to eat themselves to death, let them."

Though Key's theory gained sanction from the American Heart Association, other researchers still questioned it with conflicting ideas of what causes coronary disease. As one Philadelphia physician said, every time you question this man Keys, he says, 'I've got 5000 cases. How many do you have?"

Key's "diet-heart hypothesis" was based on his epidemiological Seven Countries study, which found an association, not a cause, between saturated fat intake, blood cholesterol, and heart disease.

His doctrine that saturated fats were the leading cause of coronary heart disease was supported by opinionated and even spurious data. It was also ignorant of the vast confounding factor of trans-fatty acids and minimised evidence against sugar, tobacco and refined carbohydrates. It concentrated its attack on fresh and less processed animal foods.

The fact that, as seen earlier, humans had been eating animal meat for nigh on 2.6 million years, or the fact that CHD had only manifested itself in the previous 50 years, obviously never occurred to him.

Around the same period, in 1957, John Yudkin, then one of the UK's leading nutritionists and founder of the nutrition department at Queen's College, University of London, published an article entitled "Diet and Coronary Thrombosis: Hypothesis and Fact" in the *Lancet Journal*.[17] The central thrust of his article was that international comparisons do not support the view that total fat or animal fat is the direct and single cause of coronary thrombosis. He stated, "There is a better relationship with sugar intake than with any other nutrient we have examined".[18]

Ancel Keys knew that Yudkin's sugar hypothesis posed an alternative to his own. If Yudkin published a paper, Keys would denounce it and him. He called Yudkin's theory "a mountain of nonsense" and accused him of issuing propaganda on behalf of the meat and dairy industries. Yudkin never responded in kind. He was a mild-mannered man, unskilled in the art of political combat.

He retired from his post at Queen Elizabeth College in 1971 to write his book, "Pure, White and Deadly," about sugar. It was published in 1972 and was written for a lay readership.[19]

It intended to summarise the evidence that the consumption of sugar was leading to a significantly increased incidence of coronary thrombosis; that it was undoubtedly involved in dental caries, probably involved in obesity, diabetes and liver disease, and possibly involved in gout, dyspepsia and some cancers.

An ominous quote from page 3 of his book stated, "If only a small fraction of what is already known about the effects of sugar were to be revealed about any other material used as a food additive, that material would promptly be banned." At the time, the British Sugar Bureau dismissed Yudkin's claims about sugar as "emotional assertions," the World Sugar Research Organisation called his book "science fiction."

In 1967, chaired by Senator George McGovern, the United States Senate Select Committee on Nutrition and Human Needs was established to address a growing concern about hunger and malnutrition in the United States. In 1974, McGovern expanded the committee's scope to include national nutrition policy. The committee shifted focus from obtaining adequate nutrients by not eating enough to avoiding excessive food components linked to chronic disease. As Ted Kennedy summarised in his opening remarks to the committee, "Although the children of West Africa melt away from starvation, America stands in ironic contrast as a land of overindulged and excessively fed. In many ways, the wellbeing of the overfed is as threatened as the undernourished".[20]

In January 1977, the committee published its first recommendations: to increase carbohydrate consumption and reduce overall fat consumption from approximately 40 to 30 percent of energy intake, reduce saturated fat consumption to 10 percent of total energy intake, and minimise sugar and salt consumption. As in the previous margarine wars, the agricultural lobbies, including those representing the meat, egg, dairy, and sugar industries, strongly objected.

Indeed, in the 1960s and 1970s, the sugar industry actively advertised sugar as an appetite suppressant and sponsored research programs to cast doubt on the hazards of sucrose, while promoting fat as the primary dietary culprit in coronary heart disease (CHD).

Figure 10.1 shows a reconstruction of the type of advertisement that featured in various popular magazines of the time, including Time and National Geographic. The PR campaign, paid for by the Sugar industry, was driven by Carl Byor & associates and was so successful that it won the prestigious "Silver Anvil" award for excellence in "the forging of public opinion".

The same ads were eventually stopped by the Federal Trade Commission, which accused the sugar industry of false

Figure 10.1 Reconstructed Sugar information advertisement from the 70's encouraging people to eat cookies so they can lose weight. Photo credit: Maria Kovalets, Unsplash.

advertising. While the sugar association issued a statement defending the advertisements as offering "sound advice to the American public."

David Stroud, CEO of the National Livestock and Meat Board, contested the scientific basis for the Dietary Goals recommendations. He argued, "Much of the poor advice has come from zealots with a good deal to say but little to no scientific evidence supporting their positions". As a concession to the meat board, a revised guideline, version two, was published 10 months later, changing the wording from "decrease meat consumption" to "choose meats, poultry, and fish which will reduce saturated fat intake".[21]

With the publication of the Dietary Guidelines, the American Society for Clinical Nutritionists, the American Heart Association (AHA), and the National Cancer Institute aligned with the low-fat recommendations. A scientific consensus was emerging that a low-fat diet was needed to prevent the two leading causes of death: coronary heart disease and cancer.

Finally, in 1980, closely followed by the UK in 1983, the first US dietary guidelines were published, advising all to eat less of something: specifically, to cut back on saturated fats and high-cholesterol foods. Of course, an increase in carbohydrates was the unavoidable consequence of the demonisation of fat. These guidelines now shaped the diets of hundreds of millions of people. Doctors advised on them, and food companies developed products to comply with them.

Sugary soft drinks now provided an alternative to traditional non-alcoholic meal accompaniments such as full-cream fresh milk. Between 1975 and 2014, sales of whole-fat milk decreased by nearly 61 percent, while low-fat and skimmed milk sales increased by almost 170 percent and 156 percent, respectively, according to USDA figures.

The consumption of soft drinks in America increased fivefold from 10 gallons (37.9 L) per person per year in 1950 to just more than 50 gallons (189.3 L) per person per year by the end of the century, equivalent to approximately one 16-oz soft drink per person per day.[22]

From a total sugar consumption perspective, in 1850, the average American consumed 20 pounds (9.1 kg) of sugar per capita. Almost a century and a half later, that same consumption had increased fivefold to 120 pounds (54.4 kg) per capita in 1994. By the early 21st century, it exceeded 160 pounds (72.6 kg) per capita.[23]

In fairness to the 1980 dietary guidelines, more than likely, the kinds of carbs the authors of the guidelines had in mind were eating more whole grains, fruits and vegetables. They also actually recommended cutting down on sugar intake. However, the message was lost in translation, and the new mantra became fat is bad; carbs are good. Reducing total fat intake led to an increase in the consumption of more refined carbohydrates and less healthy fats, and both of these changes have had significant negative consequences and global health impacts.

Frank Hu, professor of nutrition and epidemiology at the Harvard School of Public Health, sums it up: "The country's big low-fat message backfired. The overemphasis on reducing fat caused the consumption of carbohydrates and sugar in our diets to soar".[24]

Another problem with the 'all fat is bad' message is that not all fats are bad. Indeed, some dietary fats are very good, and the simplistic idea that eating fat makes one fat is invalid.

As we continue to discover, dietary fats are more than just dense calories; they constitute essential building blocks for cell membranes that ultimately govern blood clotting, heart susceptibility to erratic rhythms, LDL/HDL ratios, and central nervous system function, including intelligence and mental health.

Unfortunately, highly processed carbohydrates now comprise the unhealthiest component of the food supply. Sugar and starchy foods, such as white bread, white rice, potato products, crackers, and biscuits, digest quickly into glucose, raising insulin levels and programming the body for excessive weight gain and an increased risk of chronic disease.

Since the 1980s, when experts started advising people to eat less fat—based on the belief that a high-fat diet led to CHD—obesity has skyrocketed, creating a significant impact on national economies, reducing productivity and life expectancy and increasing disability and healthcare costs.

As we shall see later, the worldwide prevalence of obesity nearly tripled between 1975 and 2016. In 2016, more than 1.9 billion adults aged 18 years and older were overweight, including over 650 million obese adults. In the United States, 36.5 percent of adults are obese, and another 32.5 percent of American adults are overweight. Similar stats for the UK show that approximately 60 percent of the population is overweight.

Obesity rates are now linked to more deaths worldwide than underweight. Globally, more people are obese than underweight – this occurs in every region except parts of sub-Saharan Africa and Asia.

It was four decades after their 1961 cover article on Ancel Keys that Time magazine, in June 2014, published a new cover article entitled "Eat Butter. Scientists labelled fat the enemy; why they were wrong."[25]

The article states, "The low-fat results are in; the experiment was a failure. We cut the fat, but by almost every measure, Americans are sicker than ever". "The prevalence of Type 2 diabetes has increased by 166 percent from 1980 to 2012, and an estimated 86 million people are now prediabetic".

It goes on to detail that deaths from heart disease have fallen– a fact that many experts attribute to better emergency care, less smoking and widespread use of cholesterol-controlling drugs like statins–but cardiovascular disease still remains the country's No. 1 killer".

10.3 "THE TIMES THEY ARE A CHANGING"

"The Times They Are A-Changing," We Are Going on a Revolution," and "Something in the Air" are all song titles from the late sixties, Bob Dylan, the Beatles, and Pete Townshend's creation, Thunderclap Newman, respectively. The titles summarise what is now known as the counterculture movement of the 1960s and early 1970s.

Collectively known as hippies, this new generation appeared to reject the work ethic and morality of their parents' generation. The "baby boomers" who had readily adapted to post-war consumerism, such as automobiles, TVs and TV dinners![26] The movement's experimentalism, whether in drugs, art, or fashion, was symptomatic of rejecting conventional wisdom. Its espousal of love and peace also rejected Cold War diplomacy, mutually assured destruction, and, in America, the Vietnam War.

Motivated by a perfect storm of political, social, and environmental strife, the counterculture movement also took on the food industry, bringing the movement's broader political and ecological agenda to the table. The counter cuisine of hippie foods was characterised by the oppositional language of … natural *vs.* plastic, white *vs.* brown, processed *vs.* unprocessed, fast *vs.* slow, light *vs.* heavy. Counterculture-oriented publications like the Whole Earth Catalogue and The Mother Earth News became popular.

Today's health foods—from granola to tofu—as well as recycling and organic agriculture, the greater availability of ethnic cuisine, and more widespread concern for ecology, green living, and global warming all have roots in the same era. As Johnathon Landers details in his book "Hippie Food: How Back-to-the-Landers, Longhairs, and Revolutionaries Changed the Way We Eat" Young Americans wanted to strip their cuisine back to its pre-industrial roots. This "new" cuisine incorporated whole grains and legumes, organic and fresh vegetables, soy foods such as tofu and tempeh, and nutritional boosters like wheat germ and sprouted grains.[27]

It also incorporated flavours from Eastern European, Asian, and Latin American cuisines. The food these young bohemians concocted with all these ingredients was often vegetarian, sometimes macrobiotic, and occasionally inedible!

In the 1960s and early 1970s, counterculture adopted practices such as recycling and organic farming long before they became mainstream. Environmentalist global organizations such as Greenpeace and the Worldwide Fund for Nature (WWF) arose in these years.

The publication of Rachel Carson's book Silent Spring in 1962 seeded critical new ideas in the public mind: Spraying chemicals to control insect populations can also kill birds that feed on dead or dying insects. Chemicals travel not only through the environment but also through food chains.[28]

These were ideas new to the public consciousness, paramount among them was the notion that life is much more interconnected and interdependent than people assume or understand. The book outlined how we allowed chemicals to be used with little or no advance investigation of their effect on soil, water, wildlife, and man himself. It stressed that future generations are unlikely to condone the lack of prudent concern for the integrity of the natural world that supports all life.

Silent Spring was met with fierce opposition by chemical companies. Still, it spurred a reversal in national pesticide policy, leading to a nationwide ban on DDT for agricultural use. It also inspired an environmental movement that led to the creation of the US Environmental Protection Agency.

The first Earth Day in 1970 brought environmental concerns to the forefront of youth culture. Earth Day—April 22—marks the anniversary of the birth of the modern environmental movement that started in 1970 when 20 million Americans took to the streets, parks, and auditoriums to demonstrate for a healthy, sustainable environment in massive coast-to-coast rallies.[29] Groups that had been fighting against oil spills, polluting factories and power plants, raw sewage, toxic dumps, pesticides, freeways, and the extinction of wildlife suddenly realised they shared common values.

While the counterculture movement and its optimistic idealism died down in the 1970s, the seeds of sustainability and environmental awareness were planted as participants grew older and moved into mainstream society. Likewise, some of that same idealism started to mainstream in business; think of

Ben and Gerry's; "Peace, love and Ice Cream", Richard Branston; "Screw it, let's do it" or Anita Roddick and the Body Shop's "profits-with-a-principle".

As Rudy M. Baum, Editor-in-Chief of Chemical & Engineering News, published in their June 4, 2007, edition:[30] "At a time when humans largely believed themselves to be apart from nature and destined to control it, Carson argued passionately that nature is, in fact, a network of interconnections and interdependencies and that humans are a part of that network and threaten its cohesion at their own peril".

Suppose one defines the counterculture era as a complete rejection of the 250 years of the Industrial Revolution and its consequences. In that case, it is ironic that this same movement and geographic area played a significant role in the upcoming technological revolution.

The two Steves—Steve Wozniak and Steve Jobs, the developers of Apple—along with many other early computing and networking pioneers from Silicon Valley, discovered LSD and roamed the campuses of UC Berkeley, Stanford, and MIT in the late 1960s and early 1970s. From this group of social "misfits", they would go on to shape the modern world.[31]

REFERENCES

1. H. Ritchie, V. Samborska and M. Roser, Urbanization Our world in Data, https://ourworldindata.org/urbanization.
2. R. Lepkoff, P. E. Dans and S. Wasserman, *Life on the Lower East Side*, Princeton Architectural Press, 2006.
3. D. Blum, *The Poison Squad*, Penguin, 2018.
4. J. Baker, *The Law's Two Bodies*, Oxford University Press, 2001.
5. E. Khosrova, *Butter: a rich history*, Algonquin Books Of Chapel Hill, Chapel Hill, North Carolina, 2017.
6. N.Y Times, On This day, August 7th 1886, https://archive.nytimes.com/www.nytimes.com/learning/general/onthisday/harp/0807.html.
7. M. G. Venkatesh Mannar and R. F. Hurrell, *Food Fortification in a Globalized World*, Saint Louis Elsevier Science & Technology Ann Arbor, Michigan Proquest, 2018.
8. M. Saltmarsh, *Essential Guide to Food Additives*, Royal Society of Chemistry, 2013.

9. D. Mudgil and S. Barak, *Functional Foods: Sources and Health Benefits*, Scientific Publishers, 2017.
10. M. Zhang, *et al.*, "English Disease": Historical Notes on Rickets, the Bone-Lung Link and Child Neglect Issues-*Nutrients*, 2016, **8**(11), 722.
11. World Health Organization, *Understanding the Codex Alimentarius*, Food & Agriculture Orgnazation, 2018.
12. D. L. Goetsch and S. Davis, *Understanding and Implementing ISO 9000 and ISO Standards*, 1998.
13. R. Rupp, *The Butter Wars: When Margarine Was Pink*, National Geographic, 2014.
14. States. U. How Tobacco Smoke Causes Disease, 2010.
15. A. Muto, *The Early Years, 1893–1953*, University of California Press, 1993.
16. Time Magazine Ancel Keys Jan. 13th 1961.
17. J. Yudkin, Diet and coronary thrombosis hypothesis and fact, *Lancet*, 1957, **273**(6987), 155–162. Available from: https://pubmed.ncbi.nlm.nih.gov/13450357/.
18. I. Leslie, The sugar conspiracy, Guardian Newspaper, 2016.
19. J. Yudkin and R. H. Lustig, *Pure, white and deadly: how sugar is killing us and what we can do to stop it*, Penguin, London, 2012.
20. E. Klein, *Ted Kennedy*, Crown Archetype, 2009.
21. M. Gerald and I. Daniel Benrubi, McGovern's Senate Select Committee on Nutrition and Human Needs *versus* the meat industry on the diet-heart question (1976–1977), *Am. J. Public Health*, 2014, **104**(1), 59–69.
22. L. Messinger, Could low-fat be worse for you than whole milk?, The Guardian, 9th Oct 2015.
23. S. W. Mintz, *Sweetness and Power: The Place of Sugar in Modern History*, Viking, New York, NY, 1985.
24. F. Hu, *Obesity Epidemiology*, Oxford University Press, 2008.
25. B. Walsh, Eat Butter. Scientists labelled fat the enemy. Why they were wrong, Time Magazine, 23rd June 2014.
26. G. Kosc, C. Juncker, S. Monteith and B. Waldschmidt-Nelson, *The Transatlantic Sixties Europe and the United States in the Counterculture Decade*, Bielefeld Transcript Verlag, 2013.
27. J. Kauffman, *Hippie food: how back-to-the-landers, longhairs, and revolutionaries changed the way we eat*, William Morrow, An Imprint Of Harpercollins Publishers, New York, 2019.

28. R. Carson, *Silent Spring*, Penguin Books, London, 1962.
29. Earthday Organisation, https://www.earthday.org/about-us/.
30. R. Baum and R. Carson, *Chem. Eng. News*, 2007, **85**(23), 5.
31. J. Markoff, *What the Dormouse Said How the Sixties Counterculture Shaped the Personal Computer Industry*, Penguin Publishing, 2005.

CHAPTER 11

The 21st Century: Diet Food, Profitability, and Environmental Concerns

11.1 DIET FOOD

While the counterculture food movement of the early 70s was oriented around the idea that we needed to avoid industrialised foods, a future concept that may yet see its day, it had no real impact on the food industry at the time.

By the 1980s, the low-fat food approach had become an overarching ideology promoted by physicians, the US and UK governments, and popular health media. Although the food industry initially worried about a low-fat approach, by the late 1980s, food producers realised that low-fat products could provide additional selling lines and profit-making opportunities.

Fat-free yoghurt, cookies, and other processed foods—the formula was simple: replace the fat with sugar or the new cheaper alternative, High-Fructose Corn Syrup. These new low-fat industrial foods, driven by widespread advertising, discounted supermarket specials, and consumer demand, intensified in the 1980s and 1990s and continue to do so to this day.

From a formulation or cost perspective, when fat is removed from a food recipe, the same ingredient, with all its inherent

---

Food and Us: The incredible story of how food shapes humanity
By Seamus Higgins
© Seamus Higgins 2025
Published by the Royal Society of Chemistry, www.rsc.org

taste and mouthfeel characteristics, must be replaced with a similar substitute or a product that can mimic the features of the original component. From a food industry perspective, the race was on to develop new fat replacers and sweeteners derived from various food sources, including carbohydrates, protein, and vegetable oil.

Starch-based carbohydrates such as corn, soy, and other cereals or legumes such as peas were chemically transformed into modified starches or enzymatically produced as maltodextrins. High fructose corn syrup (HFCS), also made from refined cornstarch, was widely adopted by food formulators, as its usage increased by more than 1000 percent between 1970 and 1990, far exceeding the changes in intake of any other food or food group.[1]

Proteins, such as whey protein or egg whites, were used as gelling, foaming, or emulsifying agents. Vegetable oils were hydrogenated to create more solid and cheaper fats.

Today, more land is devoted to growing vegetable oil crops than all fruits, vegetables, legumes, nuts, roots and tubers combined. Vegetable oil crops are also 2 of the top 3 drivers of global deforestation.[2]

While single scientific disciplines, such as chemistry or biology, had been applied to food production from the mid-19th century onwards, instead of thinking of themselves as chemists, physicists, microbiologists, chemical engineers, *etc.*, who happened to be working in food, they were increasingly describing themselves as food scientists or food technologists.

The International Union of Food Science and Technology was officially inaugurated in Washington, DC, in 1970. In 1980, the Institute of Food Science and Technology became a full member of the UK Science Council.[3]

To cover both bases, the food industry also bought into what the Guardian newspaper has called the ultimate oxymoron, "diet food" – something you eat to lose weight![4]

When highly processed diet meals emerged, the new low-fat trend squared a seemingly impossible circle. Often created with more sugar and/or higher HFSS calorie counts than the original products, they were now marketed for weight loss "as part of a calorie-controlled diet."

Weightwatchers, created by New York housewife Jean Nidetch in the early 1960s, was bought by Heinz in 1978. Heinz, in turn, sold the company in 1999 to investment firm Artal for $735 m. Another was Slimfast, a liquid meal replacement invented by chemist and entrepreneur Danny Abraham, which was bought in 2000 by Unilever.

In June 2006, Nestlé acquired the US diet phenomenon Jenny Craig in a transaction valued at approximately $600 million. It was a clever play on words, as evidenced by the headline "Nestle sells most of Jenny Craig in slimming drive," when Reuters announced in November 2013 that Nestle had sold its weight loss business to a US private equity firm.

This event occurred just two years after Nestlé was listed as No. 1 in the Fortune Global 500 as the world's most profitable corporation, with a $200 billion market capitalisation.

11.2 PROFITABILITY AND A CHANGING MARKET

And so, to more recent times, with Heraclites' universal change constant, still very much to the fore. The problem, or opportunity, depending on how you view it, is that the same rate of change is now on steroids, increasing from a linear model to an exponential pace.

The UK population has risen from 38.3 million in 1901 to 65.1 million in 2015,[5] emulating an even more significant population growth in the USA from 76 million (1900) to 321 million in 2015.[6] Globally, the population has quadrupled from 1.7 billion in 1900 to its current count of 8 billion people.[7] The urban-rural divide in the UK is now 96 percent to 4 percent, respectively, being slightly higher than Europe and the USA's average of just over 80 percent. However, that figure is growing globally from 56 percent (4.4 billion inhabitants) to an estimated 70 percent by 2050.[8]

Over the past 50 years, global gross agricultural output has more than tripled in volume, and productivity growth in agriculture has enabled food to become more abundant and cheaper.[9] For example, the relative cost of food in the UK, as measured by the share of household budget spent on food, has dropped to just 10.7 percent or approximately 1/3 of what it was 50 years ago.[10] Hopefully, the recent global food, fuel, and fertiliser price increases primarily driven by the fallout from the war in Ukraine and sanctions imposed on Russia

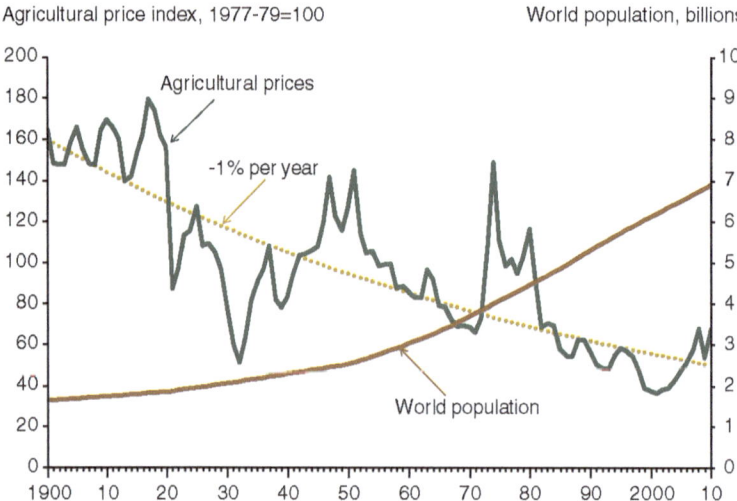

Figure 11.1 Real agricultural prices and world population 1900–2010.[9] Credit: US Department of Agriculture. Source: USDA, Economic Research Service using Fuglie, Wang, and Ball (2012). Depicted in the chart is the Grill–Yang agricultural price index adjusted for inflation by the U.S. Gross Domestic Product implicit price index. The Grill–Yang price index is a composite of 18 crop and livestock prices, each weighted by its share of global agricultural trade (Pfaffenzeller et al., 2007). World population estimates are from the United Nations.

will become temporary spikes on the same trend lines (Figure 11.1).

Along with the vast strides made in enhanced food safety, variety, and meal convenience over the last century, in keeping with population growth, the food industry has now grown to become the largest manufacturing industry in both the UK and Europe.

Planet Tracker estimated the enterprise value of the global food system to be around USD 14 trillion with revenues in the region of USD 15 to 19 trillion.[11] This is equivalent to between 16 and 20 percent of GDP.

Today's global industry now represents more than 10 percent of consumer spending and 40 percent of employment. It is also forecast to grow at a compound annual growth rate (CAGR) of 5.0 percent over the next ten years.[12]

Increasing disposable income due to economic growth, the rise of global middle-class consumers, and growing brand consciousness have also driven the food and grocery retail market in developing countries, such as India and China. All of these factors, among others, have contributed to the creation of one of the world's largest manufacturing industries, both in terms of turnover and profitability.

Suppose one looks at US stock market returns by economic sector (1963–2014). With a 13.3 per cent CAGR return, consumer staples have had the highest return of all economic sectors, outperforming the Energy, Financial, Industrial and Technology sectors.[13] A $1000 investment made in 1963 would now be worth $1 000 000 plus.

Like most businesses that have evolved since the 1970s, food companies are evaluated by investors based on their potential earnings, brand equity, and revenue growth, rather than their raw material sourcing or products produced. When investing in a food company, the bottom line is traditionally the most critical factor. You want to know how much the company earns and whether it can boost revenue streams. Is the company on a growth trajectory or in decline? These are all key factors in determining the worth of the company and your shareholding.

Supporters of food stocks argue that, regardless of economic fluctuations, companies that sell food are often among the most likely to generate a profit, as food is among the few recession-resistant goods. People always have to eat!

Consolidation has also become a growing trend in the food industry, with many of the world's larger food companies expanding their market share through mergers and acquisitions. For example, the Kraft Heinz Company was formed in 2015 through the merger of Kraft Foods Group and H.J. Heinz Company. Nestle has acquired numerous food and beverage companies, including Gerber, Purina, and Blue Bottle Coffee.

Critics of consolidation argue that it can lead to reduced competition, higher consumer prices, and job losses as redundant positions are eliminated. Proponents say that mergers can lead to increased efficiency, innovation, and profitability for shareholders.

Today's modern food industry is now a complex, global collective of diverse profit-driven businesses that supply most of the food consumed by the world's population. With 43 of the top 50 (March 2019) food companies now professionally managed "by way of the modern art," MBA or financial/economic specialists.[14] Economies of scale have been achieved through acquisitions, mergers, vertical and horizontal integration across the supply chain, and multinational operations.

The drive for increased economies of scale continues, accompanied by rising agricultural productivity and a push to reduce raw material input costs. Similarly, the increased processing of the same raw materials, health and sustainability issues aside, has led to lower cost, increased palatability, and convenience food for consumers.

Many major food companies maintain research and development departments, yet their spending in this area is the lowest among significant manufacturing sectors.[15] In contrast to companies like Amazon, Google (Alphabet), or Volkswagen, which allocate billions or a significant percentage of their profits to R&D, most food companies invest less than 3 percent of their profits back into research.

When examining a global list of the top 100 companies by research expenditure, the first food company, Nestlé, appears at position 77, with a research expenditure of only 2.35 percent. This raises questions about whether "Big Food" is adequately prepared for upcoming industry change.

The present rate of change will intensify further with what is now referred to as Industrial 4.0. The Fourth Industrial Revolution (4IR) represents the next phase in digitisation, driven by disruptive technologies, including the rise of data and connectivity, artificial intelligence and analytics, human–machine interaction, and advancements in robotics.

While 4IR will make products and services more easily accessible and transmissible, food companies, consumers, and stakeholders throughout the value chain will now have full access to the same information. Whereas critical drivers for food growth were previously led solely by taste, price, and convenience, new emerging drivers will now include health and wellness, safety, social impact, and experience, with transparency becoming an overarching driver across all these factors.

The industry is also witnessing the emergence of several new niche food brands and smaller, more focused regional companies leveraging new technology to develop innovative food production systems, including plant-based foods, insect protein, cell-cultured meat, 3D-printed foods, and personalised foods.

The last decade has seen the top 25 international food manufacturers retain a compound annual growth rate of just 1.8 percent. In contrast, small- and medium-sized companies with more specialised consumer offerings saw their sales grow with a compound annual growth rate of 11 to 15 percent.

The response of the existing "big food" players to change has been mixed. On the one hand, considerable discussion has been held about resetting company objectives to include sustainability and environmental considerations. A quoted report from the Ellen MacArthur Foundation was partly sponsored by two major players from the industry, namely, Danone and Tetrapak.

On the other hand, larger food companies are establishing venture capital funds to invest in start-ups. Nine of the ten "big food" stalwarts, including Kellogg's, Campbell Soup, and Heinz, have established new venture capital companies to "more fully participate" in growth opportunities.[15]

Although the predominant economic model of the commercial food system, as will be seen in the next section, has been poorly aligned with social, health, environmental, and sustainability goals, recognition of this challenge and an appetite for change are slowly emerging, driven by more health-conscious and publicly aware consumers.

Oxfam's "Behind the Brands" campaign is a recent example of a concerted effort to address social and environmental concerns on a global scale. The same campaign brought public pressure on the "Big 10" food companies—Coca-Cola, Danone, General Mills, Kellogg, Mars, Mondelez, Nestlé, PepsiCo, Unilever, and Associated British Foods (ABF)—to improve their social and environmental policies.[16]

The campaign focused on agricultural sourcing from developing countries, examining workers, smallholders, and women, as well as the impact on communities and assessing awareness, knowledge, commitments, and supply chain management.

The emphasis on transparency has since galvanised targeted companies to put far more information on their websites. According to Oxfam's campaign, they aim for a more equitable global food value chain. Moving away from current business models founded on short-term profit maximisation towards more holistic business models that internalise social and environmental performance and good governance.

When Oxfam launched its campaign in 2013, it used a single, incredibly informative image to detail and highlight the world's largest food and beverage companies, their leading brands, and the myriad of product brands they own, control, and produce.

The same image can be viewed at: https://www.oxfamamerica.org/explore/stories/these-10-companies-make-a-lot-of-the-food-we-buy-heres-how-we-made-them-better/.

Oxfam has granted us permission to reprint their image in the book. However, due to copyright and intellectual property issues, we are unable to include the image here.

A more recent campaign from an aptly named organisation called Bite Back is "Fuel Us, Don't Fool Us". Their first research report in a series conducted in partnership with the University of Oxford investigated the 10 biggest global food and drink businesses operating in the UK and their sales of packaged food and drink products. Their report also highlights the mission statements of some of these companies (Figure 11.2).

"We believe the world we want tomorrow starts with how we do business today"; **Mars Inc.**

"We are Nestle, the good food, good life company. We believe in the power of food to enhance lives"; **Nestle SA.**

"Our mission: Create more smiles with every sip and every bite... Nourishing products... Conserving nature's precious resources and fostering a more sustainable planet for our children and grandchildren". **PepsiCo Inc.**

We are driven by our purpose: to make sustainable living commonplace. We want to do more good for our planet and our society—not just less harm. We want to act on the social and environmental issues facing the world and enhance people's lives with our products. **Unilever Group.**

Diet Food, Profitability, and Environmental Concerns 137

Summary of businesses' UK sales from packaged food and drinks

	Ferrero & related parties	Mondelez International Inc	Unilever Group	Kellogg Co	Mars Inc	Nestlé SA	PepsiCo Inc	Coca-Cola Co, The	Kraft Heinz Co	Danone, Groupe	**Totals**
Number of brands included in the analysis	19	40	26	14	28	42	28	23	10	11	**241**
Number of products included in the analysis	347	965	648	149	346	641	768	530	515	389	**5,298**

Ferrero & related parties	Mondelez International Inc	Unilever Group	Kellogg Co	Mars Inc
100%	98%	84%	77%	72%
£919.3 mn	£2820.4 mn	£1256.1 mn	£777.2 mn	£1458.0 mn

Nestlé SA	PepsiCo Inc	Coca-Cola Co, The	Kraft Heinz Co	Danone, Groupe
70%	68%	36%	33%	2%
£1252.1 mn	£2095.2 mn	£1086.6 mn	£391.8 mn	£27.4 mn

■ % of sales from HFSS   £ Estimated value of sales from HFSS (£mn)

Figure 11.2 UK sales from packaged food and drinks. Bite Back/University of Oxford.[17] Source: Bite Back, *Fuel us, don't fool us: Are food giants rigging the system against children's health?*, 2024.

11.3 ENVIRONMENTAL AND SUSTAINABILITY CONCERNS

The Gaia hypothesis, named after the ancient Greek goddess of the Earth, posits that the Earth and its biological systems behave as a single, living, colossal entity. Developed by scientist James Lovelock and microbiologist Lynn Margulis in the 1970s, the Gaia hypothesis proposes that life, through its interactions with the Earth's landmass, oceans, and atmosphere, produces a stabilising effect on conditions on the planet's surface conducive to all its life forms.[18]

The Gaia hypothesis is considered controversial. Scientists say it conflicts with conventional scientific thinking and fails to explain many aspects of the Earth's physical and biological systems. Topics related to the hypothesis include how the biosphere and the evolution of organisms affect the stability of global temperature, the salinity of seawater, atmospheric oxygen levels, the maintenance of a hydrosphere of liquid water, and other environmental variables that affect Earth's habitability.

The theory behind the hypothesis is similar to how our body's biology has evolved over millions of years to develop its own principle of Homeostasis, from the Greek words for "same" and "steady". In effect, creating an automated process by which the body's biological systems maintain stability while adjusting to changing external conditions.

Maintaining homeostasis requires the human body to monitor its internal conditions continuously relative to its external environment. Each physiological state has a specific set point, ranging from body temperature to blood pressure, as well as particular nutrient and blood sugar levels. It employs various homeostatic mechanisms to sustain its optimal functioning. Indeed, the human body would be unable to function proficiently if there was a prolonged imbalance in internal physical conditions and chemical composition.[19] Indeed, as discussed later, could a highly processed diet significantly contribute to that same chemical imbalance?

While the current global food system has supported a fast-growing population and fuelled economic development and urbanisation, these productivity gains have come at a considerable environmental cost. Our global food system is now the most significant contributor to biodiversity loss, deforestation, drought, freshwater pollution and the collapse of aquatic wildlife.

The Agri-Food sector is now the world's second-largest emitter of greenhouse gases after energy, responsible for nearly one-third of all human-caused emissions.[20] A recent report from the Ellen MacArthur Foundation entitled "Cities and the Circular Economy for Food" puts a price on it; for every dollar spent on food, society now pays two dollars in health, environmental, and economic costs.[21]

A third of all edible food produced goes uneaten, despite 10 percent of the global population going hungry. The rise in overnutrition is leading to obesity and other diet-related chronic diseases in many developing countries.

Pesticides, synthetic fertilisers used in conventional farming practices, and manure mismanagement in factory farming exacerbate air pollution and contaminated soils and can leach chemicals into water supplies. The poor management of food waste and by-products generated during food processing, distribution, and packaging further pollutes water, particularly in emerging economies.

Poor agricultural practices significantly contribute to the 39 million hectares of soil degraded each year globally.[22] Approximately 70 percent of global freshwater demand is required to sustain farming practices. Large-scale commercial agriculture and local subsistence agriculture were responsible for about 73 percent of deforestation between 2000 and 2010.[23]

The world now relies on just three crops for more than 50 percent of its plant-based protein, contributing to a dramatic loss of biodiversity (over 60 percent in the last 40 years), increased vulnerability to diseases and pests, and greater reliance on chemical inputs.[24]

The global mass of farm animals is now 22 times more than all wild animals combined.[25]

From an environmental perspective, James Lovelock, co-author of the Gaia Hypothesis, passed away last year (2024) at the age of 103. In one of his last interviews, he explained that when he suggested 60 years ago that our planet was self-regulating like a living organism, he was roundly criticised by academic scientists and considered an outsider as an independent scientist.

In the years since, we have seen how much life—especially human life—can affect the environment. He emphasised that two genocidal acts—suffocation by greenhouse gases and the clearance of the rainforests—have caused changes on a scale not seen in millions of years.

"My fellow humans must learn to live in partnership with the Earth; otherwise, the rest of creation will, as part of Gaia, unconsciously move the Earth to a new state in which humans may no longer be welcome. The recent COVID-19 pandemic may have been a negative feedback loop. Gaia will try harder next time"[26]

"Our global food system is broken", says Paul Polman, former global CEO of Unilever. "We realise that we can't have infinite growth on a finite planet, and the costs of not acting become higher than the costs of acting. We need to move to a net-positive food system and create a reliable food value chain for the long term".[27]

REFERENCES

1. G. A. Bray, *et al.*, Consumption of high-fructose corn syrup in beverages may play a role in the epidemic of obesity, *Am. J. Clin. Nutr.*, 2004, **79**(4), 537–543.
2. J. Nobbs, The Environmental Impact of Vegetable Oils, https://www.jeffnobbs.com/posts/the-environmental-impact-of-vegetable-oils.
3. IFST, Our background, https://www.ifst.org/about-ifst/our-background.
4. J. Peretti, Fat profits: how the food industry cashed in on obesity, The Guardian Newspaper, 2013.
5. Office for National Statistics, Overview of the UK population: Gov UK, March 2017.
6. United States Census Bureau, Historical Population Change Data (1910–2020), 2021.
7. H. Ritchie, L. Rodés-Guirao, E. Mathieu, M. Gerber, E. Ortiz-Ospina, J. Hasell and E. Rosser, *Population Growth*, Our World in Data, 2023.
8. H. Ritchie, V. Samborska and M. Roser, *Urbanisation*, Our World in Data, 2024.
9. S. Wang and K. Fuglie, *New Evidence Points to Robust But Uneven Productivity Growth in Global Agriculture*, USDA Economic Research Service, 2012.
10. UK Department for Environment, Food & Rural Affairs Food statistics in your pocket, UK Gov., 2022.
11. Planet Tracker, Valuing the Global Food System, 2023, https://planet-tracker.org/what-is-your-food-worth/.
12. J. Pomeroy, *The future of food – can we meet the needs of 9bn people*, HSBC Insights, 2023.
13. J. Charalambakis, *Returns by Economic Sector (1963–2014)*, Black Summit Financial Group Inc., 2020.
14. D. Heft and S. Higgins, Food industry and engineering—Quo vadis?, *J. Food Process Eng.*, 2021, **44**(8), e13766.
15. M. Geller, *Food revolution: Megabrands turn to small start-ups for big ideas*, Reuters, 2017.

16. Oxfam, The Behind the Brands Campaign 2013–2016.
17. Biteback, Fuel us don't fool us, https://www.biteback2030.com/the-gut-punch/the-fuel-us-dont-fool-us-report/.
18. L. Margulis and D. Sagan, *Slanted truths*, Copernicus, 1997.
19. S. Libretti and Y. Puckett, in *Physiology, Homeostasis*, StatPearls Publishing, StatPearls Treasure Island (FL), 2025, [updated 2023 May 1].
20. M. Crippa, E. Solazzo and D. Guizzardi, *et al.*, Food systems are responsible for a third of global anthropogenic GHG emissions, *Nat. Food*, 2021, **2**, 198–209.
21. D. Heft and S. Higgins, *Re-engineering the food Industry; where do we go from here Applied Degree Education and the Future of Learning*, Springer Nature Singapore, Singapore, 2022.
22. S. Scherr, *Soil Degradation A Threat to Developing-country Food Security by 2020?* International Food Policy Research Institute, 1999.
23. FAO, *Global Forest Resources Assessment*, Food and Agriculture Organization of The United Nations, Rome, 2018.
24. A. M. D. Ortiz, C. L. Outhwaite, C. Dalin and T. Newbold, A review of the interactions between biodiversity, agriculture, climate change, and international trade: research and policy priorities, *One Earth*, 2021, **4**(1), 88–101.
25. R. Mc Kie, 'A wake-up call': total weight of wild mammals less than 10% of humanity's, The Guardian Newspaper, 18th March 2023.
26. J. Lovelock, Beware: Gaia may destroy humans before we destroy the Earth, The Guardian Newspaper, 2nd Nov. 2021.
27. P. Polman and A. Winston, *Net Positive: How Courageous Companies Thrive by Giving More Than They Take*, Harvard Business Review Press, USA, 2021.

CHAPTER 12

Food and Us: Our Unique DNA and Physical Makeup

When we compare our current biological species, *Homo sapiens*, to Australopithecus Afarensis, our Lucy, the early hominin that lived around 3.17 million years ago, we have evolved physically into an entirely different being regarding appearance, cognitive abilities, and lifestyle—evolution as defined as a gradual or heritable change in the genetics of a population over time. As we have also seen, human evolution is a continuous process, and we continue to evolve with each new generation.

The Eagle, located on Bennett Street in Cambridge, is a 17th-century UK pub, one of the university town's best-known drinking establishments. In February 1953, a jubilant Francis Crick walked into this Grade II-listed building and proclaimed he and James Watson had "found the secret of life."

Watson and Crick went on to share a portion of the 1962 Nobel Prize in Physiology or Medicine for their theories about DNA structure and replication, as presented in "A Structure for Deoxyribose Nucleic Acid" (DNA) and Its Genetic Implications".[1]

The discovery of DNA revolutionised evolutionary biology, and DNA analysis provided new answers to several old questions, particularly those related to Darwin's theory of evolution and heritable change. Darwin's observations determined that natural

---

Food and Us: The incredible story of how food shapes humanity
By Seamus Higgins
© Seamus Higgins 2025
Published by the Royal Society of Chemistry, www.rsc.org

populations contain a wealth of variation, and he saw how these variants accumulate through natural selection. However, he did not know how variations physically occurred or how these traits could be inherited.

The discovery of DNA's structure, its double helix design, and ease of replication laid the groundwork for several of today's complex genetic questions and research. It has also given us a clearer picture of how our bodies have evolved physically and how our current food system mismatches our energy requirements.

In 2003, the Human Genome Project (HGP), a landmark global scientific effort, produced a genome sequence that accounted for over 92 percent of the human genome.[2] (All our genes together are known as our "genome").

The HGP revealed that humans have approximately 25 000 genes. Genes are segments of your DNA that give you the physical characteristics that make you unique. They could be considered blueprints or instruction manuals for your body's operation.[3]

Our genes are found in all 37 trillion cells that comprise our body and direct specific bodily processes by coding for proteins. From building a particular body part to determining our physical characteristics or traits such as hair colour, eye colour, skin tone, and more. Genes also encode proteins that indirectly support bodily functions. These help the immune system respond to injury or assist blood flow through the circulatory system.

Your genome is inherited from both of your parents. During meiosis, which is the process the body uses to produce egg and sperm cells, chromosomes (the packages of DNA) replicate themselves but with half the number of chromosomes as the original. This is how you receive your genetic material equally from each parent.

Like your genome, each gamete (reproductive cell) is unique, which explains why siblings from the same parents may not look alike. Unless you are an identical twin. Identical twins share the same genomes and are always of the same sex. They are created from the fertilisation of a single egg by a single sperm, with the fertilised egg then splitting into two.

Genetic mutations may occur during cell division when cells and their DNA divide and replicate. A mutation is a change in an organism's DNA sequence.[4]

Mutations can result from errors in DNA replication during cell division, exposure to mutagens, or a viral infection.

Mutations that occur in eggs and sperm can be passed on to offspring, whereas mutations that occur in body cells are not. As we have seen, some mutations can confer a selective advantage on an individual and become more common over time. Ultimately, these mutant genes may drive the older versions out of existence.[5]

If one reviews some of our more recent physical evolutionary changes from a time perspective, the chance of Steve Earl or Ed Sheeran meeting their blue-eyed Galway girl 6000 years ago could never have happened! Initially, we all had brown eyes and a genetic mutation in a single individual in Europe 6000 to 10 000 years ago led to the development of blue eyes.

From a food perspective, 5000 years ago, virtually no adult human could properly digest animal milk. However, through evolutionary change, northern Europeans began inheriting a genetic mutation that enabled them to do so.

More recently, researchers have found another genetic variant in populations in India and East Asia that have favoured vegetarian diets over many generations. This new genetic variant enables these individuals to efficiently metabolize omega-3 and omega-6 fatty acids from a plant-based diet.[6]

Further gene research identified the mutation of the HERC2 gene, which results in blue eyes. Another mutation in the SLC24A5 gene, which lightens skin colour, is now found in up to 95 percent of Europeans. As per the above examples and the mutations that have emerged in one population or another, lactose intolerance (the inability to digest lactose, a component of animal milk) is caused by variants in the LCT gene.[7]

The vegetarian genes or alleles found in India and East Asia appear to control the FADS1 and FADS2 enzymes in the body. These enzymes are crucial for converting omega-3 and omega-6 fatty acids into what researchers refer to as "downstream products," which are essential for brain development and the control of inflammation.[8] All of the above examples are inherited gene mutations.

Although the human genome contains approximately 25 000 genes, it only comprises about 2 percent of our DNA sequence. Over 98 percent of our DNA that does not code for proteins was initially referred to as "junk DNA." In other words, the function of the remaining 98 percent of the genome, which is non-coding, was not yet fully understood!

More recently, scientists have discovered that, despite its name, "junk DNA" also plays a critical role in how our bodies behave. After sequencing the genomes of thousands of patients suffering from various genetic disorders, scientists have discovered that 88 percent of changes to the genetic code that correlated with disease were found in the "Junk DNA."

In September 2003, The ENCyclopedia Of DNA Elements (ENCODE) project was established as an international research effort funded by the National Human Genome Research Institute (NHGRI). It was introduced to facilitate the identification and analysis of the complete functional elements in the human genome sequence.[9]

The ENCODE project represents the first systematic effort to determine the location and organisation of all functional elements. This new data and research reinforce our knowledge that the genome contains minimal unused sequences and is a complex, interwoven network. What has also become apparent is that researchers now recognize that these non-coding regions have critical biological functions that remain to be determined.

12.1 GENOTYPES AND PHENOTYPES

A person's genotype is their unique sequence of DNA. More specifically, this term refers to the two forms or alleles a person has inherited from their parents for a particular gene. A phenotype is the observable expression of a genotype in a person's physical appearance.

A phenotype is a term used to describe the range of observable characteristics that can vary from one individual to another. These characteristics can include height, eye or hair colour, or the presence or absence of disease. While genetics plays a role in determining certain phenotypic traits, environmental factors can also significantly influence them. In some cases, environmental factors may be even more important than genetics in determining a particular phenotype.[10]

Phenotypic variation in humans is a direct consequence of genetic variation, which interacts with environmental and behavioural factors to produce phenotypic diversity.

Food components can alter phenotypes by interacting with genes and changing their expression profiles. Thus, a gene can disturb an individual's response to their diet by affecting its absorption, metabolism, or nutrient transportation.[11]

For example, gene variants that promote the practical storage of dietary sugar in the form of fat tissue as an energy reserve had a natural advantage for survival when sweet things were only available during a short harvest period, such as a surplus of fruits. However, because sweet foods are permanently available in modern society, the influence of genetic variation on health can often only be assessed in the context of the relevant environmental conditions.

This mismatch between our ancient physiology and a modern diet underlies many of the so-called diseases of civilisation we experience today, such as coronary heart disease, obesity, hypertension, type 2 diabetes, epithelial cell cancers, auto-immune disease, and osteoporosis [12] By comparison, the same diseases are rare or virtually non-existent in still-living hunter-gatherer tribes and non-westernised populations.

There is also growing evidence that an addictive-eating phenotype may exist.[13] Hence, the more recent significant debate regarding whether highly processed or "Ultra" processed foods (UPFs) with refined carbohydrates or added fats are addictive. UPFs that can trigger compulsive use and higher dietary intake are positively associated with binge eating, even in the face of significant diet-related health consequences (*e.g.* diabetes and other metabolic diseases). The combination of refined carbohydrates (such as sugar) and fat appears to have a supra-additive effect on human reward encoding. As we will see in a later section, UPFs and their components increase dopamine in the striatum at a similar magnitude as nicotine when delivered orally (150–200 percent).

Just like tobacco, they also trigger compulsive use and intense cravings, and their continued use is highly reinforcing. Importantly, if the science supports that UPFs are unhealthy and addictive, this challenges the assertion that excessive UPF intake is purely a matter of choice. Likewise, the surge in addictive, ultra-processed, energy-dense, and low-nutrient food consumption and marketing has an adverse impact on the health of individuals, societies, and the broader ecosystems on which we depend.

12.2 OUR MICROBIOME

If genes are the basic unit of heredity passed from parent to child, mitochondria govern a cell's energy production and other metabolic functions. Offspring inherit mitochondria from their mother because of mitochondrial DNA.[14]

Our mitochondrial DNA accounts for a small portion of our total DNA, comprising only 37 of the body's approximately 25 000 protein-coding genes. Mitochondria are organelles that produce ATP (adenosine triphosphate), the energy source used and stored at a cellular level. All living cells rely on ATP for energy. Without it, we could not form a thought or move a muscle.

ATP is produced through cellular respiration, which takes place in a cell's mitochondria. Mitochondria are tiny subunits within a cell that specialise in extracting energy from the foods we eat and converting it to ATP. The microbiome is, in essence, comprised of microbes: microbial cells that number in the trillions and reside within us. They have helped us to evolve, shaped our biology, and defined the success of our species.[15]

The discovery that gut microbes evolved alongside their human hosts offers another perspective on the human gut microbiome. Gut microbes have been passed between people over hundreds to thousands of generations, so as humans have changed, so have their gut microbes.

Scientists have been studying the microbiome for over 30 years, but there is still a lack of understanding about its diversity and its role in human health and disease. However, our understanding of the link between the human microbiome and diseases such as obesity, inflammatory bowel disease, arthritis and autism is rapidly expanding. Despite the progress, many knowledge gaps still need to be filled.[16]

What is known is that the microbiome communicates directly with our mitochondria, which is crucial for cell and organ function. This community of microorganisms works with our mitochondria and is essential in digestion and absorption, vitamin production, immune regulation, and metabolic and neurological health.

We also know that newborns get their first microbiome from their mothers during birth. The baby is completely covered with bacteria during a natural birth, giving it a brand-new microbiome.

This early bacteria bath also helps regulate early gut colonisation, immune maturation, and infant brain development.[17]

Studies have also revealed that gut microbiome composition is established within the first three years of life. Influencing factors include delivery mode, feeding, and many other factors, such as environmental (including geographic location) and household influences, rather than genetics.

As we age, the immune system and gut microbiome undergo significant changes in composition and function that correlate with increased susceptibility to infectious diseases and reduced vaccine responses. The more recent COVID-19 pandemic illustrated this fact, where age and frailty became the strongest predictors of morbidity and mortality from the virus infection.

It is also widely recognized that our dietary choices have a significant impact on the gut microbiome, affecting its overall structure, composition, and functionality. This complex system of microorganisms resides within the gut. It interacts with the gut epithelium, a single layer of cells that lines the digestive tract and mucosal immune system, playing an integral role in maintaining a healthy digestive environment. By promoting intestinal homeostasis, the gut microbiome helps maintain our bodies' optimal functioning and supports our overall well-being (Figure 12.1).[18]

If you have ever "gone with your gut" to make a decision or felt "butterflies in your stomach" when nervous, you are likely getting signals from your second brain.[20]

A better understanding of the microbiome in our digestive system now often referred to as the "brain in your gut," has revolutionised medicine's knowledge of the links between digestion, mood, health, and how you think. Scientists refer to this "little brain" as the enteric nervous system (ENS).[21]

The gut–brain axis (GBA) is a bidirectional link between the central nervous system (CNS) and the enteric nervous system (ENS) of the body. It involves direct and indirect pathways between cognitive and emotional centres in the brain, as well as peripheral intestinal functions.

As a result, our microbiome impacts everything—and I mean everything: it's critical to every part of our health, affecting brain function, mood, fertility, susceptibility to weight gain, illness, allergies, how we respond to cancer treatment and how fast we age.[15]

Food and Us: Our Unique DNA and Physical Makeup 149

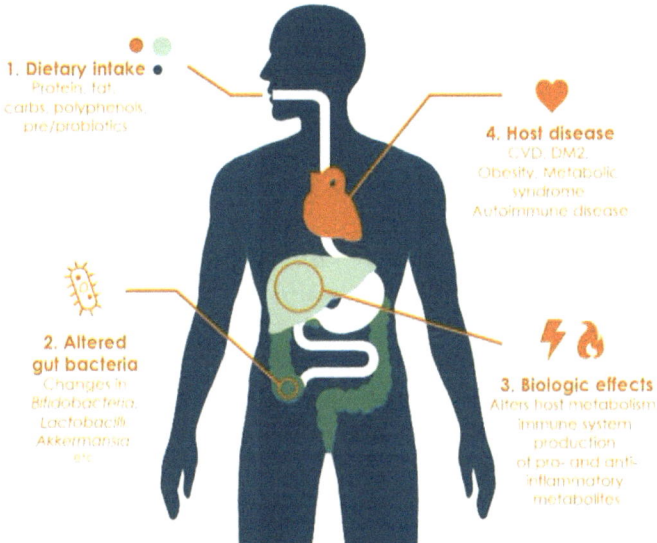

Figure 12.1 Influence of diet on the gut microbiome and implications for human health.[19] Reproduced from ref. 19, https://doi.org/10.1186/s12967-017-1175-y, under the terms of the CC BY 4.0 license, http://creativecommons.org/licenses/by/4.0/.

Of course, the evolution of the human diet, from natural food sources to ultra-processed foodstuffs,[22] has also led to drastic changes in the gut microbiota. These changes have negatively affected immune signalling in the intestines and the brain.

For example, the loss of a high-fibre diet mainly affects the type and amount of microbiota in the intestines. Evidence has shown that soluble dietary fibre slows gastric emptying, increases perceived satiety and plays a significant role in appetite regulation.[23]

Dietary fibre is broken down and fermented only by enzymes from the microbiota in the large intestine. As a result of this fermentation process, short-chain fatty acids (SCFAs) are produced, which lower the pH of the colon. (acidity level). This decrease in pH helps to reduce the growth of harmful bacteria and stimulates the activity of immune cells.

When it comes to food and how our body works, as Giulia Enders outlines in her book Gut: The Inside Story of Our Body's Most Underrated Organ,[24] the essential factor is not to reduce the human body to a two-dimensional cause-and-effect machine.

The brain, the rest of the body, its bacteria, and the elements in our food all interact in four dimensions.

12.3 OUR FOOD AND OTHER ANIMALS

The human genome contains approximately 3 billion DNA building blocks that make us living things. Most cells in the body are diploid, with 23 pairs of chromosomes, which results in approximately 6 billion base pairs of DNA per cell. Only a handful are uniquely ours. In fact, despite our outward differences, humans are 99.9 percent genetically similar.[25]

All people's DNA base pair sequence is nearly identical—that is what makes us human. However, slight differences in the order of the six billion base pairs in everyone's DNA cause variations in hair colour, eye colour, nose shape, and so on. Hence, no two people, other than identical twins, have precisely the same DNA sequence.[26]

Most of the differences we notice are caused by a tiny fraction of DNA. Given six billion base pairs per cell, that tiny fraction—1/1000 of six billion base pairs—is still six million different base pairs per cell. Therefore, there is ample room for genetic differences among us.

Although we differ in a tiny proportion of our DNA, we can still vary by many DNA bases.

But how alike are we to other, non-human life forms? It turns out we are a lot more similar than you might think. This is because large chunks of our genome perform similar functions across the animal kingdom. It turns out that all mammals, including ourselves, are descended from an ancestral species that lived about one hundred million years ago. As a result, we share roughly 90 per cent of our DNA with cats, dogs, cattle, and elephants.[27]

Indeed, elephants resemble humans in several ways: consider their massive brains, empathetic social bonds, lengthy gestations, high intelligence, and long life spans. Moreover, their offspring require an extended period of dependent care.

Mice share approximately 97.5 percent of their working DNA with humans, and genetically engineered mice have become indispensable for investigating the genetics of diseases and testing new drugs.

Humans, chimpanzees, and gorillas also have highly similar DNA sequences, as would be expected given that we are all part of the same family, Hominidae. Humans and chimpanzees have 98.6 percent of their DNA in common, while humans and gorillas have just slightly less, at 98.3 percent.[28] So, how do animals with similar DNA and evolutionary histories balance their food requirements regarding food choice?

The big lesson we can learn from animals is not what they eat but how they eat. Animals possess five appetites: protein, carbohydrates, fat, salt, and calcium, all in that order. A key message from studying animal food behaviour is that the appetite for protein is the strongest for many animals and can override all other cravings.

In "Eat like the Animals", Professors David Raubenheimer and Stephen Simpson explore what nature can teach us about the science of healthy eating.[29]

The authors make a compelling case for their "protein leverage hypothesis." Their studies have shown that, with only minor exceptions, all animals leverage protein because of their inherent protein requirements. They will unconsciously eat to meet those requirements. If the only foods available are low-protein, they will overeat to get as close as possible to their protein requirement.

Beyond understanding animal foraging in general, this has significant implications for understanding human obesity. Many animal species will overeat calories if given low-protein diets to satisfy their protein needs—hence the protein-leverage hypothesis. In natural food environments, these appetites work together to help animals select a balanced diet.

While humans possess this ability, the modern food environment has become so altered that our appetites can no longer function in harmony. Instead, they compete, each vying for its nutrients. This competition causes us to overeat fats and carbohydrates. As the authors point out, we've diluted protein in the food supply with ultra-processed fats and carbs. We have also disconnected the brake on our appetite systems by decreasing dietary fibre. This is perfect for encouraging us to eat and buy more, but it is detrimental to our health.

Ultra-processed foods (UPFs) that increasingly dominate human diets are generally low in protein and are often crafted to have an extra sweet or savoury taste, including umami—thus mimicking natural protein requirements. "Surround yourself with whole foods, such as nuts, fruits, vegetables, healthy oils, unrefined grains, pulses, and moderate amounts of quality meats," says Professor Raubenheimer. "That way, we can expose the amazing appetite systems we share with other species to a food environment where they can work their magic and lead us to a balanced diet".

REFERENCES

1. J. D. Watson, A. J. Berry and K. Davies, *DNA: the story of the genetic revolution*, Alfred A. Knopf, New York, 2017.
2. National Research Council (U.S.), *Committee On Mapping And Sequencing The Human Genome. Mapping and sequencing the human genome*, National Academy Press, Washington, D.C., 1988.
3. R. Plomin, *Blueprint: how DNA makes us who we are*, The Mit Press, Cambridge, Massachusetts, London England, 2019.
4. R. Dawkins, *The Selfish Gene*, Oxford University Press, Oxford, 1976.
5. B. Bogin, *Patterns of human growth*, Cambridge University Press, Cambridge, United Kingdom, New York, NY, 2020.
6. K. Ramanujan, Eating green could be in your genes, Cornell Chronicle, 29 March 2016.
7. J. Parrington, *The deeper genome: why there is more to the human genome than meets the eye*, Oxford University Press, Oxford, United Kingdom, New York, NY, 2017.
8. T. D. Gelehrter, F. S. Collins and D. Ginsburg, *Principles of medical genetics*, Williams & Wilkins, Baltimore, 1998.
9. O. P. Rajora, *Population Genomics: Concepts, Approaches and Applications*, Springer International Publishing, Cham, 2019.
10. G. O'Brien and W. Yule, *Behavioural Phenotypes*, Cambridge University Press, 1995.
11. Food Forum, *Nutrigenomics and the Future of Nutrition*, National Academies Press, 2018.
12. P. Roberts, *The Impulse Society: America in the Age of Instant Gratification*, Bloomsbury, New York, NY, 2015.
13. M. D. Lewis, *The biology of desire: why addiction is not a disease*, PublicAffairs, New York, 2015.

14. N. Lane, *Power, Sex, Suicide Mitochondria and the Meaning of Life*, Oxford University Press, Michigan Proquest, Oxford, Ann Arbor, 2018.
15. Food Forum and Institute of Medicine, *The Human Microbiome, Diet, and Health*, National Academies Press, 2013.
16. A. Fasano and S. Flaherty, *Gut feelings: the microbiome and our health*, The Mit Press, Cambridge, Massachusetts, 2021.
17. T. Harman and A. Wakeford, *Your Baby's Microbiome*, Chelsea Green Publishing, 2017.
18. M. Lyte and J. F. Cryan, *Microbial Endocrinology: The Microbiota-Gut-Brain Axis in Health and Disease*, Springer, New York, NY, 2014.
19. R. Singh, *et al.*, Influence of diet on the gut microbiome and implications for human health, *J. Transl. Med.*, 2017, **15**(1), 73.
20. G. Tsipursky, *Never Go With Your Gut*, Red Wheel/Weiser, 2019.
21. M. D. Gershon, *The second brain: a groundbreaking new understanding of nervous disorders of the stomach and intestine*, Harper Perennial, 2019.
22. P. S. Ungar and M. F. Teaford, *Human diet*, Bergin & Garvey, Westport, Conn., 2002.
23. C. M. Galanakis, *Dietary fiber: properties, recovery, and applications*, Academic Press, 2020.
24. G. Enders, *Gut: the Inside Story of Our Body's Most Underrated Organ (Revised Edition)*, Greystone Books, 2018.
25. B. Alberts, A. Johnson and J. Lewis, *Molecular Biology of the Cell*, Garland, New York, 4th edn, 2002.
26. J. D. Watson and A. Berry, *DNA*, Knopf, 2009.
27. R. Dawkins and Y. Wong, *The Ancestor's Tale*, Weidenfeld & Nicolson, 2016.
28. J. Marks, *What it means to be 98 percent chimpanzee: apes, people, and their genes*, University Of California Press, Berkeley, 2003.
29. D. Raubenheimer and S. J. Simpson, *5 Appetites: Eat Like the Animals for a Naturally Healthy Diet*, HarperCollins UK, 2020.

CHAPTER 13

Divining Food Energy and Nutritional Intake

13.1 PROCESSED FOODS AND ADDITIVES

As we have seen, humans have engaged in various food-processing techniques throughout history. Cooking, for instance, enhanced food safety and improved digestibility and palatability. The advent of milling and baking marked a significant milestone in creating a new staple food industry. Similarly, the preservation and processing of food laid the foundation for what we now recognise as the food processing industry.

The 19th century saw a significant leap in food processing technology with the introduction of pasteurisation and canning. These new techniques revolutionised both food safety and preservation methods.

As we have also explored, the 20th century brought about a notable shift in the purpose of food processing, transitioning beyond just food safety and preservation methods to the mass production of convenience foods.

The popularity of convenience foods has driven the development of more advanced food processing technologies, including spray drying, freeze-drying, extrusion, hydrogenation, irradiation, high-pressure processing, and pulsed electric field treatment.

Food and Us: The incredible story of how food shapes humanity
By Seamus Higgins
© Seamus Higgins 2025
Published by the Royal Society of Chemistry, www.rsc.org

The advent of food science specialists also added to the mix, literally, with new food additives and modified food processing techniques such as caramelisation, coagulation, denaturation, emulsification, gelatinisation and the Maillard reaction.

Today, the US Food and Drug Administration (FDA) has approved over 10 000 additives to preserve, package, or modify the taste, appearance, texture, or nutritional content of foods.[1]

Salt, sugar, and corn syrup are by far the most widely used additives in food in the United States.

In Europe, these additives are referred to as E numbers. The European Food Safety Authority (EFSA) has authorised more than 300 substances as food additives.[2]

Some substances used as additives occur naturally, such as vitamin C (E 300) and pectin (E 440) in fruit, lycopene (E 160d(ii)) in tomatoes, and lecithin (E 322), which is present in a range of foods, such as egg yolks, soya beans, peanuts, and maize.

Other food additives, such as synthetic lycopene or E 160d(i), can now be chemically synthesised. They can also be derived from animals, insects (*e.g.*, carminic acid or E 120, obtained from cochineal insects), or minerals (*e.g.*, calcium carbonate or E 170, obtained from ground limestone).

Likewise, food packaging has become more sophisticated and varied with the development of modern materials and technologies. Advanced techniques, such as vacuum packaging (for meat) and modified atmosphere packaging (for crisps), are now widely used in the industry to extend shelf life while maintaining freshness and mouthfeel.

All of these changes in the modern food system have become part of our regular diets and have ushered in a new era of various extended shelf life and more palatable food types.

The problem is that, in the past half-century, a different type of food processing has been developed, says Fernanda Rauber, a nutritional epidemiologist at the University of São Paulo, Brazil, about what we now call "ultra-processed foods" (UPF). "These substances would not be found in our kitchen. Usually, they contain little to no proportion of real foods."[3]

But before we consider what could be regarded as UPF, let us examine the food labels that currently detail the nutritional and energy content of the food we eat. These labels, now a legislated requirement for most food types, are duly displayed by food manufacturers and service providers on food packaging or other media, such as websites and menu boards.

Until the late 1960s, food labels provided little or no information about the nutrient content of our food. It wasn't until the passage of the Nutrition Labelling and Education Act of 1990 that mandatory nutrition labelling—and the introduction of the Nutrition Facts panel that we know today—was expanded to include all processed foods regulated by the FDA in the United States.[4]

The United Kingdom introduced a similar act in 1996 with stated Guideline Daily Amounts. This system was also adopted in the European Union, and similar systems were replicated in several other countries.

Dr Louis W. Sullivan, then Secretary of the US Department of Health and Human Services, directed the FDA to undertake a comprehensive initiative to revise the Food Labeling Act, stating that as consumers shop for healthier food, they often encounter confusion and frustration. The grocery store has become a Tower of Babel, and consumers must be linguists, scientists and mind readers to understand the many labels they see.

According to the FDA, understanding the Nutrition Facts label on food items can help individuals make healthier food choices. The label breaks down the number of calories, carbs, fat, fibre, protein, and vitamins per serving of the food, making it easier to compare the nutrition of similar products.[5]

But given what we know about the complexity of our bodies, our individuality, our DNA, our food digestive system and brain control linkages, is a one-size-fits-all engineering definition of energy, initially defined for steam engines, an appropriate or indicative measure of the nutritional value of the food we eat today?

13.2 FROM STEAM ENGINES TO BOMB CALORIMETERS AND FOOD ENERGY LABELS

It was a French chemistry professor, Nicolas Clement, who sought a word to describe how steam engines converted heat into work. In 1819, he defined the word as a Calorie for his lectures.[6] From the Latin word calor (heat), he used the phrase calorie to define a simple measurement of heat energy, *i.e.* the amount of energy needed to raise 1 kilogram of water 1 degree Celsius at standard atmospheric pressure.

Another French chemist and physicist, Pierre Berthelot, created the bomb calorimeter in the 1870s. He designed the device to measure the amount of heat absorbed or released during chemical or physical changes, such as combustion, to determine the specific heat capacity of various organic materials, including food. It is still used today in various applications, including measuring the energy content of fuels, food, and other materials.

The bomb calorimeter's working mechanism involves burning organic material or a food sample in an oxygen-rich environment and measuring the resulting temperature increase. The energy released by the combustion reaction is then applied to heat a known amount of water. The water temperature change (ΔT) is then utilised to calculate the sample's energy content.[7] As per Clements's above calorie definition and the first law of thermodynamics, its calorific value can then be determined. Ask any engineer or physicist, and they will tell you that the first law of thermodynamics states that the total energy of a system remains constant, even if it is converted from one form to another.[8]

However, bomb calorimeters aren't the only method for measuring food calories. Food scientists also rely on calculations developed by the 19th-century US chemist Wilbur Atwater, who determined an indirect way to estimate the number of calories in food products.[9]

Atwater introduced his 4-9-4 system, having developed his own "respiration calorimeter." He believed a bomb calorimeter could not account for humans losing some calories through heat or passing urine and faeces. In his experiments, he fed different foods to human volunteers and then measured the heat of combustion of their faeces. By calculating the difference in the heat of combustion between the food eaten and the subsequent waste material passed, he could estimate the calories the volunteers absorbed.

Atwater presented his calculations to the world in 1900. He determined that we absorb 9 kcal per gram of fat, 4 kcal per gram of carbohydrates, and 4 kcal per gram of protein from food. Hence the name, the 4-9-4 system. He also found that alcohol has seven calories per gram.

One kilocalorie (kcal) is equivalent to 1 large Calorie (Cal) or 1000 small calories (cal). The terms "kcal" and "Calories" (with a capital "C") are often used interchangeably in everyday contexts, particularly when discussing food energy and calorie counts.

Every calorie count on every food label you have seen is based on these estimates or modest derivations of the same calculation. To complicate matters further, the US FDA and other regulatory authorities also allow a margin of error of up to 20 percent for the stated calorie value displayed compared to the actual value of energy consumed.[10]

Over the years, the calorie-counting model has been associated with many imperfections. Not to mention that the human body does not behave like a steam engine!

The "calories in, calories out" formula for weight loss success is a myth because it oversimplifies the complex process of calculating energy intake and expenditure.[11] More importantly, it fails to consider the mechanisms our bodies trigger to counteract a reduction in energy intake. Likewise, it does not consider other factors, such as the gut microbiome, metabolic adaptation, hunger hormone regulation, or the fact that two people could eat the same food and absorb different amounts of calories.

In addition to energy values, a food package's front, back, and sides are filled with details, which are now a legal requirement in most countries to inform consumers about the food's ingredients, additives, preservatives, allergens, manufacturer, and other relevant information. Most ingredient lists are also required to quantify or list the ingredients used in the product in descending order, according to the percentage ratio of each ingredient.

There are various types of front-of-pack nutrition labels worldwide. The UK uses a colour-coded system called a traffic light labelling system. Unlike the US, energy values are displayed per 100-gram or pack-size portions, and the red, amber, and green colours provide an at-a-glance view of the fat, saturated fats, sugars, and salt levels in a portion of a food or drink. Red for high, green for low.

The EU uses a reference intake table that also displays the energy contribution in the product relative to one's recommended daily diet allowance. However, the RIs only show the amounts of harmful nutrients a product contains, such as fat, sugars, and salt, and do not include positive nutrients, such as fibre.[12]

In 2017, France introduced its Nutri-Score label, which has also been adopted in Belgium, Spain, and Portugal.[13] The system is determined by assessing healthy and unhealthy nutrients. It considers the contents of energy, simple sugars, saturated fat, and sodium (as harmful components of the product), as well as the percentage of fruit and vegetables, and the amounts of fibre and protein (as positive components). The scheme, however, does not help consumers identify specific unhealthy ingredients, such as salt or sugar.

While there is no universally agreed-upon definition of processed or "ultra-processed" foods yet, some new classification systems have been developed that attempt to group foods by their level of processing.

A recent UK SACN (Scientific Advisory Committee on Nutrition) report on processed foods and health reviewed existing classifications on food processing and a review of available evidence on the association between processed food consumption and health outcomes.[14]

The NOVA food classification system was the only one identified that met all five of the committee's screening criteria. While the observed associations between higher consumption of (ultra-) processed foods and adverse health outcomes were reported "as concerning," their overall conclusion was, as would be expected from a committee of 22 academics, "there are uncertainties around the quality of evidence available" as such, "this subject needs more research".

According to the British Nutritional Institute, the term 'ultra-processed foods' is based on NOVA, a food classification method that categorises foods according to how they are prepared and the extent of processing they have undergone. They define Ultra-processed foods as formulations of ingredients, mostly of exclusive industrial use, typically created by a series of industrial techniques and processes. Examples include sweets, crisps, sweetened-flavoured yoghurt, manufactured bread, breakfast cereals, and packaged soups.[15]

As per the British Nutritional Institute, "Several research studies that have made headlines in recent years have linked the consumption of ultra-processed foods to negative health outcomes such as obesity, cancer, and mortality.... Factors explicitly linked to processing, such as additives, process contaminants, and compounds formed at high temperatures, may be contributing to the observed effects. The design of these studies does not allow us to draw conclusions about any cause-and-effect relationships between ultra-processed food consumption. It makes it difficult to determine which properties of the foods included in the 'ultra-processed' category could be responsible for the observed associations".

However, just as in the butter margarine wars, as already outlined earlier, given the above statement, one must also consider the vested interests of the British Nutritional Institute. In particular, the stated companies that are funding their output on ultra-processed foods.

As per page 44 of their most recent financial accounts, these companies include Arla Foods Ltd, Coca-Cola Great Britain and Ireland, General Mills, Kellogg Europe Trading Ltd, Kerry Foods Ltd, Mars UK Ltd, Nomad Foods Ltd, PepsiCo UK Ltd, Quorn (Marlow Foods Ltd) and Tate & Lyle Plc. a British-headquartered, global supplier of food and beverage products to food and industrial markets.[15]

13.3 WHY NOVA?

When Carlos Monteiro first qualified as a doctor in 1972, he worried that Brazilians weren't getting enough healthy food. By the late 2000s, as seen in many other countries, his country suffered, with the exact opposite problem.[16] In the late 2000s, after three decades of research, he could see that Brazilians were buying way less oil, sugar, and salt than they had in the past. Despite this, people were piling on the pounds. Between 1975 and 2009, the proportion of Brazilian adults who were overweight or obese more than doubled.

If people were buying less fat and sugar, why were they getting bigger? The answer he concluded was right there in the same data. Brazilians had not reduced their consumption of fat, salt, and sugar—they were consuming these nutrients in an entirely new form. People swapped traditional foods—such as rice, beans, and vegetables—for pre-packaged bread, sweets, sausages, and other processed snacks. The share of biscuits and soft drinks in Brazilians' shopping baskets had tripled and quintupled, respectively, since the first household survey was conducted in 1974.

At first glance, Monteiro's findings seemed obvious. If people overeat unhealthy food, they put on more weight. But Monteiro wasn't satisfied with this answer. He thought that something fundamental had shifted in the nation's food system. For over a century, nutritional science has focused on nutrients through calorie counting: eating less saturated fat, avoiding excess sugar, and getting enough vitamin C, among other recommendations.

Monteiro wanted a new way of categorising food that emphasised how products were made, not just what was in them. Food processing in itself was not the issue, as nowadays, practically all food is processed in some sense and to some extent.

He concluded that it wasn't just the ingredients that made a food unhealthy; it was the entire system: how the food was processed, how quickly we ate it, and how it was sold and marketed.

Monteiro created a new food classification system, as in the word Nova that breaks things down into four simple categories.

13.3.1 Group 1: Unprocessed and Minimally Processed Foods

Fruits, vegetables, and unprocessed meats are natural foods that have undergone minimal processing or alteration. These would include edible parts of plants (fruit, seeds, leaves, stems, roots, tubers) or of or from animals (muscle, fat, offal, eggs, milk), and also fungi and algae, all after separation from nature, spring and tap water.

Unprocessed foods are foods altered by industrial processes such as removing inedible or unwanted parts, drying, powdering, squeezing, crushing, grinding, fractioning, steaming, poaching, boiling, roasting, pasteurisation, chilling, freezing, placing in containers, vacuum packaging, non-alcoholic fermentation, and other methods that do not add salt, sugar, oils or fats or other food substances to the original food.

The main aim of these processes is to extend the life of unprocessed foods, enabling their storage for longer use, making them edible, and often making their preparation easier or more diverse. Infrequently, minimally processed foods contain additives that prolong product duration, protect original properties, or prevent the proliferation of food spoilage microorganisms.

13.3.2 Group 2: Processed Culinary Ingredients

These are substances derived from group 1 foods or nature by pressing, refining, grinding, milling, and drying. Some methods used to produce processed culinary ingredients are ancient in origin. But now they are usually industrial products, designed to make durable products suitable for home, restaurant and canteen kitchens to prepare, season and cook freshly prepared dishes and meals.

In isolation, processed culinary ingredients are often unbalanced, lacking in some or most nutrients. In addition to salt, they are energy-dense, containing 400 or 900 kilocalories per 100 grams. This is around 3–6 times more than cooked grains and 10–20 times more than cooked vegetables.

Oils are rarely, if ever, consumed on their own. They are combined with foods to make more palatable, diverse, nourishing, and enjoyable meals and dishes such as stews, soups and broths, salads, breads, preserves, drinks, and desserts. Thus, oils are used to cook grains (cereals), vegetables, legumes (pulses), and meat and are added to salads. Table sugar is used to prepare fruit or milk-based desserts.

13.3.3 Group 3: Processed Foods

These include canned or bottled vegetables or legumes (pulses) preserved in brine, whole fruit preserved in syrup, tinned fish preserved in oil, some types of processed animal foods such as ham, bacon, pastrami, and smoked fish, manufactured bread, and simple cheeses to which salt is added.

Processed food products usually retain the fundamental identity and most constituents of the original food. However, when excessive amounts of oil, sugar, or salt are added, they can make a dish nutritionally unbalanced. Except for canned vegetables, their energy density ranges from moderate (around 150–250 kilocalories per 100 grams for most processed meats) to high (around 300–400 kilocalories per 100 grams for most cheeses).

Processes and ingredients in this group are designed to increase the durability of group 1 & 2 foods and make them more enjoyable by modifying or enhancing their sensory qualities. They may also contain additives that prolong the product's duration, preserve its original properties, or prevent the proliferation of microorganisms.

13.3.4 Group 4: Ultra-processed Foods

Ultra-processed foods are formulations of ingredients, mostly of exclusive industrial use, typically created by a series of industrial techniques and processes (hence the wording 'ultra-processed').

Some common ultra-processed products are carbonated soft drinks; sweet, fatty or salty packaged snacks; candies (confectionery); mass-produced packaged bread and buns; cookies (biscuits); pastries, cakes and cake mixes; margarine and other spreads; sweetened breakfast 'cereals' and fruit yoghurt, 'energy' drinks; pre-prepared meat, cheese, pasta and pizza dishes; poultry and fish 'nuggets' and 'sticks'; sausages, burgers, hot dogs and other reconstituted meat products; powdered and packaged 'instant' soups, noodles and desserts, baby formula; and many other types of product.

Processes involved in the manufacture of ultra-processed foods encompass several steps and different industries. They start by fractioning whole foods into substances, including sugars, oils and fats, proteins, starches, and fibre. These substances are often obtained from a few high-yield plant foods, such as corn, wheat, soy, cane, or beet, or pre-prepared animal by-products.

Some of these substances are then subjected to hydrolysis, hydrogenation, or other chemical modifications. Industrial techniques such as extrusion, moulding, and pre-frying are used. Colours, flavours, emulsifiers, and other additives are frequently added to make the final product more palatable or hyper-palatable.

Generally, a practical way to identify if a product is ultra-processed is to check its list of ingredients for at least one item characteristic of the ultra-processed food group.

These are either food substances never or rarely used in kitchens or classes of additives whose function is to make the final product palatable or more appealing. These include hydrolysed proteins, soya protein isolate, gluten, casein, whey protein, 'mechanically separated meat', fructose, high-fructose corn syrup, 'fruit juice concentrate', invert sugar, maltodextrin, dextrose, lactose, soluble or insoluble fibre, hydrogenated or interesterified oil. The presence in the list of ingredients of one or more food substances identifies a product as ultra-processed, according to Monteiro *et al.*[17] Ultra-processed foods: what they are and how to identify them.

The processes and ingredients used to manufacture ultra-processed foods are designed to create highly profitable products characterised by low-cost ingredients, a long shelf life, and emphatic branding. These products are also convenient, ready-to-consume, and hyper-palatable, making them liable to displace freshly prepared dishes and meals from all other NOVA food groups.[18]

Is the Nova grouping of processed foods, as described above, based on scientific laws, such as the first law of Thermodynamics? The answer is no. Are the same definitions considered contentious? The answer is yes. However, the same system does provide a simplified framework for grouping edible substances based on the extent and purpose of food processing applied to them. While the view of food manufacturers has already been discussed by way of the British Nutritional Institute, the view of some academics involved in food research is not dissimilar. They believe the terms 'processing' and 'ultra-processing', which are crucial to the NOVA classification, are ill-defined and lack scientific rigor due to undefined parameters and unclear theoretical grounds. They also believe that ultimately the classification conflicts with traditional, evidence-based food evaluation methods.

Although the UK has concluded that more research is needed before using a Nova classification for processed food products, some countries have already adopted NOVA to limit ultra-processed foods in their population's dietary guidelines, for example, Brazil, Peru, Belgium, Ecuador, Israel, Maldives, and Uruguay. It has also been proposed that the U.S. Department of Agriculture (USDA) needs to address ultra-processed foods in Americans' 2025–2030 Dietary Guidelines.

Researchers at the NIH (National Institutes of Health, US) investigated whether people ate more calories when exposed to a diet composed of ultra-processed foods compared to an unprocessed diet. Despite the ultra-processed and unprocessed diets being matched daily for presented calories, sugar, fat, fibre, and macronutrients, people consumed more calories when

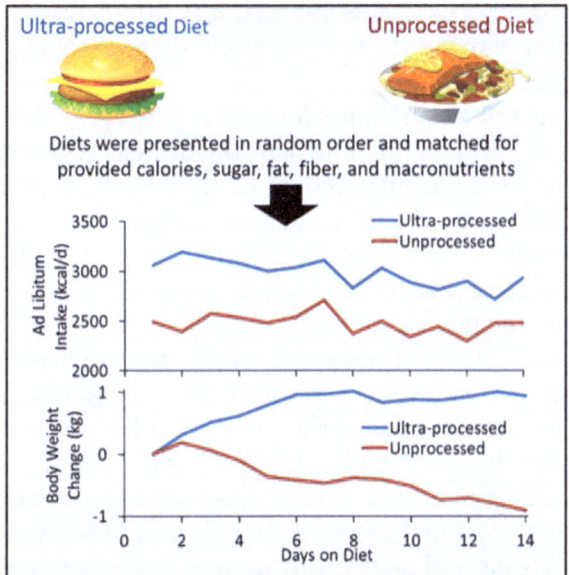

Figure 13.1 People consumed more calories when exposed to the ultra-processed diet than the unprocessed diet.[19] Reproduced from ref. 19 with permission from Elsevier, Copyright 2019.

exposed to the ultra-processed diet than the unprocessed diet. Furthermore, people gained weight on the ultra-processed diet and lost weight on the unprocessed diet (Figure 13.1).[19]

A 2024 meta-analysis report recently published in the *British Medical Journal* summarises the following concerning ultra-processed foods: hundreds of epidemiological studies and meta-analyses have reported associations between ultra-processed food consumption and adverse health outcomes. In a linked paper, Lane *et al.*[20] carefully reviewed the evidence from 45 meta-analyses encompassing almost 10 million participants. They found direct associations between exposure to ultra-processed foods and 32 health parameters, including mortality, cancer, and mental, respiratory, cardiovascular, gastrointestinal, and metabolic ill health."

13.4 DOPAMINE: THE PATHWAY TO PLEASURE

When we think of some of our more modern favourite food treats, such as milk chocolate, ice cream, chips, pizza, biscuits, buttered popcorn, or cheeseburgers, it is interesting to note that they all share similar properties, aside from their classification as ultra-processed foods.

Unlike natural whole foods found in nature, they all contain a nearly equal ratio of fat and carbohydrates, typically ranging from 1:1 to 1:2. For instance, milk chocolate contains 30 g fat and 58 g carbs, ice cream 12 g fat and 24 g carbs, crisps 30 g fat and 50 g carbs. A cheeseburger would have 14 g fat and 30 g carbs. With chocolate or ice cream, most of the carbohydrate comes from sugar.

These foods create a response within the brain that triggers the release of dopamine, a natural reward system. Similar to animal diets, as discussed earlier, the brain has separate systems to distinguish between foods that are high in fat and those that are high in carbohydrates. However, we get a resultant dopamine hit when both systems are activated simultaneously.

Human breast milk is the only natural food that deviates from this same food ratio. Breast milk contains 3–5 percent fat and 6.9 to 7.2 percent carbohydrates. It follows that people who were breastfed as babies may value foods with similar compositions as the ultimate comfort food.[21]

"It's very plausible that breast milk is somehow optimum," said Alain Dagher, a neurologist and researcher at McGill University in Montreal. "Because it's the first food that we eat, it's likely that we learn at a young age that the food is rewarding—and for the rest of our lives, that particular [fat/carbohydrate] combination becomes comfortable."

Again, as we have seen regarding food and our physical evolution, we are still only beginning to understand the symbiotic relationship between our physiology and the food we eat, just as a competitive food industry is also intent on exploiting the same weaknesses.[22]

The "bliss point" in food manufacturing refers to the perfect combination of sugar, salt, and fat that makes us crave more of a particular food. The term was first coined by Howard Moskowitz, an experimental psychologist and creator of marketing research methods in food technology. Moskowitz explained that the bliss point is the point at which a food's sugar, salt, and fat content are perfectly balanced to create maximum pleasure for our taste buds. He also wrote the book "Selling Blue Elephants: How to

Make Great Products that People Want Before They Even Know They Want Them"![23]

From a physiological perspective, it was in 1957 that Arvid Carlsson, a Swedish pharmacologist, first posited that dopamine was an independent neurotransmitter in our body's nervous system. Having devoted his life to understanding how the brain works, he was awarded the Nobel Prize in Physiology in 2000 for his research into dopamine.[24]

From an evolutionary standpoint, over millions of years, dopamine developed to reward you when you do what you must to survive—eat, drink, and reproduce. It is now understood that humans' brains are hard-wired to seek out behaviours that release dopamine in their reward systems. Food was scarce for most of evolution, and you needed this motivation to survive.

When it comes to reproduction, dopamine is activated during both sexual desire and sexual behaviour. In particular, activation of D1 and D2 dopamine receptor subtypes in the preoptic area of the hypothalamus may underlie its role in sexual desire and sexual arousal.[25]

Indeed, anything that gives you pleasure will trigger the release of dopamine. This can range from a fun activity you enjoy, like dancing, sports or cooking, to sex, shopping, and even certain drugs. Dopamine activates the reward pathway in the brain, leading you to desire these activities more.

Mounting evidence also suggests this reward-related motivation interacts with our homeostatic energy balance system. The dopamine system is critical in controlling feeding behaviour through the reward-related circuit. It contributes to the homeostatic regulation of energy balance by participating in the hypothalamic control of food intake.

Neuroscientific studies conducted using advanced techniques such as brain scanning, magnetic resonance imaging (MRI), and positron emission tomography (PET) have provided solid evidence that foods rich in fat and sugar, which are generally considered to be more palatable, can activate the dopamine reward system like tobacco or other addictive drugs. This suggests that there may be a link between the consumption of such foods and addiction-like behaviour.[26]

There is also an increasing body of research indicating that dopamine may have a significant impact on the development of obesity.[27] A decrease in the activity of the dopaminergic system, also known as dopaminergic hypofunction, can lead to excessive food intake. While genetic factors and individual phenotypes may contribute to overeating and weight gain, it is also possible that it is a compensatory mechanism of the brain to counteract the impact of excessive dopamine stimulation.

In the last twenty to thirty years, we have gained a better understanding of the genetic, neural, and environmental factors that contribute to obesity. It is clear that the current obesity crisis results from a significant disconnect between the neurobiology that drives food intake in humans and the vast array of food options made readily available by the food industry, as well as our social and economic systems.

Perhaps this deeper understanding of the same physiological cause and effect can lead to better ways of tackling the obesity pandemic by way of a multifaceted approach that involves prevention, treatment, and policy intervention.

REFERENCES

1. E. Pacifici and S. Bain, *An overview of FDA regulated products: from drugs and cosmetics to food and tobacco*, Academic Press, London, 2018.
2. M. Sadler, *Foods, Nutrients and Food Ingredients with Authorised EU Health Claims*, Woodhead Publishing, 2015.
3. G. Cediel, F. Rauber, R. Mendonça, A. Meireles, M. A. Leite and M. Gombi-Vaca, *Ultra-Processed Foods and Human and Planetary Health*, Frontiers Media SA, 2023.
4. States. U, FDA Nutrition Labeling Manual, 1993.
5. E. Wartella, A. H. Lichtenstein and C. S. Boon, Institute Of Medicine (U.S.). Committee On Examination Of Front-Of-Package Nutrition Rating Systems And Symbols, Institute Of Medicine (U.S.), Food And Nutrition Board, National Academies Press, 2010.
6. J. L. Hargrove, History of the Calorie in Nutrition, *J. Nutr.*, 2007, **136**(12), 2957–2961.
7. D. Sherwood and P. Dalby, *Modern Thermodynamics for Chemists and Biochemists*, Oxford University Press, 2018.
8. O. A. Ijabadeniyi, *Food Science and Technology*, Walter de Gruyter GmbH & Co KG, 2023.

9. A. L. Merrill and B. K. Watt, *Energy Value of Foods*, 1955.
10. B. Resnick, *Why Our Nutrition Facts Need an Overhaul*, The Atlantic, 2014.
11. G. Yeo, *Why Calories Don't Count*, Simon and Schuster, 2021.
12. European Food safety Authority, Dietary reference values, https://www.efsa.europa.eu/en/topics/topic/dietary-reference-values.
13. For O, On O. France: Country Health Profile 2021, Oecd Publishing, Paris, 2021.
14. Office for health Improvement & Disabilities, SACN statement on processed foods and health - summary report Gov. UK, 2023.
15. British Nutrition Foundation, British Nutrition Foundation position statement on the concept of ultra-processed foods (UPF), 28th May 2024.
16. M. Reynolds, Fat, Sugar, Salt … You've Been Thinking About Food All Wrong Wired, 2023, https://www.wired.com/story/ultra-processed-foods/.
17. C. A. Monteiro, *et al.*, Ultra-processed foods: what they are and how to identify them, *Public Health Nutr.*, 2019, **22**(5), 936–941.
18. F. Visioli, *et al.*, The ultra-processed foods hypothesis: a product processed well beyond the basic ingredients in the package, *Nutr. Res. Rev.*, 2023, **36**(2), 340–350.
19. K. D. Hall, *et al.*, Ultra-Processed Diets Cause Excess Calorie Intake and Weight Gain: An Inpatient Randomized Controlled Trial of Ad Libitum Food Intake, *Cell Metab.*, 2019, **30**(1), 67–77.e3.
20. M. M. Lane, E. Gamage and S. Du, *et al.*, Reasons to avoid ultra-processed foods. Ultra-processed food exposure and adverse health outcomes, *BMJ*, 2024, **384**, e077310.
21. K. Severson, This Magic Ratio Of Fat To Carbs Makes The Perfect Comfort Food, Huffpost 04/02/2019.
22. M. Moss, *Hooked*, Random House, 2021.
23. H. R. Moskowitz and A. Gofman, *Selling Blue Elephants*, Pearson Education, 2007.
24. V. K. Yeragani, *et al.*, Arvid Carlsson, and the story of dopamine, *Indian J. Psychiatry*, 2010, **52**(1), 87–88.
25. D. Z. Lieberman and M. E. Long, *The Molecule of More*, BenBella Books, 2018.

26. D. Val-Laillet, E. Aarts, B. Weber, M. Ferrari, V. Quaresima, L. E. Stoeckel, M. Alonso-Alonso, M. Audette, C. H. Malbert and E. Stice, Neuroimaging and neuromodulation approaches to study eating behaviour and prevent and treat eating disorders and obesity,, *NeuroImage: Clin.*, 2015, **8**, 1–31.
27. M. Singh, Mood, food, and obesity, *Front. Psychol.*, 2014, **5**, 925.

CHAPTER 14

Rethinking Our Food System's Status Quo

14.1 FOOD SUPPLY, MONEY AND PROFIT

As we have seen, the evolution of our food system has taken many turns and has undergone significant change over millions of years. Just as we, as humans, have evolved from Lucy, one of our original hominin ancestors, to our present-day classification as *Homo sapiens*.

Our rich history of food development and physical evolution confirms that food systems matter for everyone on this planet—they always have.

As we have seen more recently, the way we produce, market, and consume food has undergone rapid changes. This swift evolution of our food system, once deeply ingrained in our culture and social norms, has now become integrally linked with our economic, ecological, and political systems.

Food consumption is no longer influenced solely by survival needs. The range of available food products is extensive, and seasonality no longer applies. Consumers benefit from relatively low prices compared to their income, as well as high convenience foods that have accompanied changes in food production and globalisation.

Food and Us: The incredible story of how food shapes humanity
By Seamus Higgins
© Seamus Higgins 2025
Published by the Royal Society of Chemistry, www.rsc.org

On the one hand, our food system has achieved remarkable progress in keeping pace with population growth, ensuring food safety, reducing forms of malnutrition, and providing an abundant food supply to many parts of the world.

On the other hand, progress has been uneven across the world, and the way our current food system operates fuels some of humanity's most significant challenges, including hunger, undernutrition, the obesity epidemic, biodiversity loss, environmental damage, and climate change.

If one considers our evolution from hunter-gatherers, when we began to walk upright approximately 3.17 million years ago, we have only been farming for 0.003 percent of that time. We have only lived in urban areas and consumed industrially produced food for approximately 0.00003 percent of that period. Our physical evolution has been a slow process that has taken many generations of reproduction to become evident in various populations. Human traits that have emerged more recently, such as light skin, blue eyes, and lactase persistence (the ability to digest milk after weaning), have taken hundreds to thousands of years for their effects to become evident.

On the other hand, cultural change has happened at a much faster pace. Take money and our current financial system. Today, money is arguably one of the most important constructs of our recent human history. Its invention allowed our ancestors to break out of the constraints of a bartering system, and it has now become the essential medium of exchange enabling the production, distribution, and consumption of goods and services in our modern society.

As explored earlier the first use of money only appeared in Mesopotamia a few thousand years ago, around 3500 BC. Since then in a relatively short evolutionary period, money has replaced food as a symbol of power, wealth and security. From an individual perspective, money now plays a significant role in our daily lives, determining access to resources such as food, shelter, and healthcare.

From a business perspective, monetary performance has also become a de facto measure of a food company's success.[1] As explored in Chapter 7, business management, as a "new science" or "one of the modern arts," combined with the Milton Friedman doctrine of the 1970s, made maximising shareholder value the ultimate goal of a business corporation. Friedman also argued that a company has no social responsibility to the public or society. Its only responsibility is to its shareholders.

Milton Friedman's shareholder value theory has significantly impacted the corporate world.[2] It has become pervasive in the financial community and much of the business world, influencing various topics, from performance measurement and executive compensation to shareholder rights, the role of directors, and corporate responsibility.

Hence, the quandary regarding our contemporary food supply system and the role of major companies now controlling it. Given that food production is a basic necessity of life and matters to everyone, how does one reconcile humanity's primary need for a stable food supply and subsequent health and wellness with a commercial food system operating in an economic environment that demands continual profit growth?

When Henri Nestle sold his factory in 1875, a deal that included his name, signature, and bird's nest trademark, based on his family's logo, for a million Swiss francs,[3] he probably had no idea that his company would eventually become the world's largest food production company. One hundred and fifty years later, Nestlé has annual sales of over 91 billion CHF (approximately $ 103 billion), 447 factories, and a range of global food brands, including Nespresso, Nescafé, Kit Kat, Smarties, Nesquik, Vittel, and Maggi, with operations in 189 countries worldwide.

From a financial perspective, since 1970, Nestlé's share price has gone from 2.03 CHF to 97.51 CHF in 2023 (a 48-fold increase). Over the past 15 years (2009–2023), they have also managed to return 180.8 CHF billion to shareholders through share buybacks and dividends.[4]

More recently (April 2024), a coalition of shareholders representing $1.7 trillion in assets led by ShareAction filed a shareholder resolution challenging Nestlé to improve its impact on people's health and set targets to shift its sales towards healthier products. The coalition included Legal & General Investment Management (LGIM), Candriam, and the UK Parliamentary Contributory Pension Fund.[5]

While the company claims in its mission statement that its products have "the power to enhance lives," as Catherine

Howarth, Chief Executive at ShareAction, outlines, three-quarters of Nestlé's global sales consist of unhealthy products containing high levels of salt, sugar, and fats.

In response to the planned AGM resolution, Nestlé said ShareAction "is targeting the wrong company." "While we take note of ShareAction's perspective, we disagree with the notion that we should aim to limit growth in specific areas of our portfolio. A proportional target would require us to weaken valuable parts of our portfolio and create opportunities for competitors without yielding public health benefits".

The motion was duly defeated, with 88 percent voting against it, 11 percent in favour, and 1 percent abstaining.[6]

Of course, food manufacturing is only one part of our food system that has been monetised for financial gain. The entire food system has been significantly impacted due to its essential nature, market characteristics, growth potential, and opportunities for high economic returns.

The increasing involvement of financial players in commodities trading has led to greater price volatility. Speculation in agricultural commodities markets, driven by hedge funds and other financial institutions, frequently leads to significant price fluctuations that can impact producers and consumers.

Financial derivatives such as futures and options allow investors to bet on future price movements in the agricultural sector. While these instruments can provide hedging opportunities for farmers and grain purchasers, they also contribute to price increases and instability when used predominantly for speculative purposes.[7]

The Chicago Board of Trade (CBOT) is a critical player in the global agricultural commodities market, trading significant volumes of grain annually. Wheat, corn, and soybeans are among the most actively traded commodities. Corn futures alone see an average daily trading volume of around 350 000 contracts. The standard contract size of 5000 bushels (approximately 136 metric tonnes) per trade translates into billions of bushels traded annually.[8]

Money tells the story. Since the tech bubble burst in 2000, dollars invested in commodity index funds by 2013 had increased 50-fold.[9] According to Statista, the number of futures contracts traded globally has grown by 142 percent over the past decade, from 12.13 billion in 2013 to 29.32 billion in 2022. The number of traded options contracts increased from 9.42 to 54.53 billion in the same period.[10]

Commodity trading is based on the difference between the current price of a commodity in the local market and its price in a futures market. Despite the increasing financial value of futures trading, actual global grain production of wheat and corn is approximately 2000 million tons per year (corn: 1235 million tons and wheat: 765 million tons). This actual grain figure is less than 2 percent of total futures grain trading.

A recent 2023 report from the UN on trade and development highlights market concentration in critical sectors, such as the trading of agricultural commodities, as above. Likewise, how the unregulated financial activity significantly contributed to the profits of global grain traders in 2022 (Figure 14.1).[19]

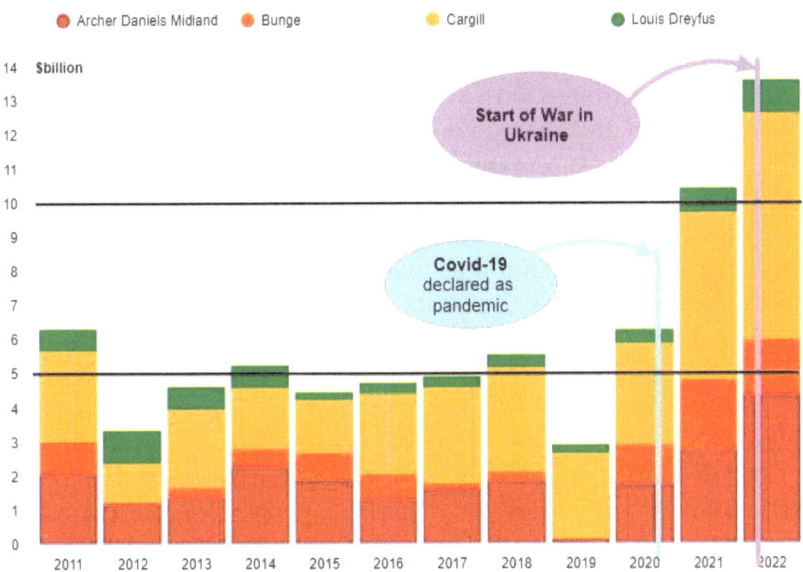

Figure 14.1 Grain trading profiteering in times of crisis. Data source: UN Trade and Development Report 2023. UNCTAD calculations based on Eikon Refinitiv and Louis Dreyfus Commodities Financial Results reports.

The more recent period of heightened price volatility since 2020, brought about by the COVID-19 pandemic and the war in Ukraine, has enabled certain major grain trading companies, in particular the ABCDs*, to earn record profits while food prices have soared globally and millions of people faced a cost-of-living crisis.

- Known as the ABCD group for the alphabetic convenience of their initials, ADM, (Archer Daniels Midland) Bunge, Cargill, and (Louis) Dreyfus account for between 75 percent and 90 percent of the global grain trade, according to estimates.

Just as non-food players have financialised grain trading, the quest for growth potential and opportunities for high financial returns has also fuelled a wave of mergers and acquisitions in the food industry. Private equity and investment funds have increasingly considered acquiring stakes in food and agriculture companies.

Warren Buffet's Berkshire Hathaway investment fund first invested $1 billion, or 37 percent of its book value, in Coca-Cola shares in 1988/9. Pleased with returns on food and drink, he arranged the largest takeover of a pure food producer with a $28 billion buyout offer for H. J. Heinz in 2013. In March 2015, 3G Capital partnered with Warren Buffett to acquire Kraft Foods for $40 billion and merged it with H. J. Heinz to form the world's fifth-largest food company.[11]

Due to his investment success, Buffett is one of the best-known investors in the world. As of June 2024, he had a net worth of $135 billion and has invested in several agricultural and food companies over the years, including, but not limited to, Archer Daniels Midland, John Deere, Monsanto, Mondelez International, Dairy Queen, Restaurant Brands International (Burgerking and Tim Hortons), The Kraft Heinz Company, and many more.

Today, the Coca-Cola Company is primarily controlled by institutional shareholders who hold 63 percent of the company. Berkshire Hathaway is the largest shareholder, with a 9.3 per cent shareholding (400 million shares) valued at $25 billion. If one had invested $1000 in Coca-Cola shares at the same time as Warren Buffett in 1988, that same investment would now be worth $23 399.50 as of June 2023, excluding dividends. The investment would have increased by a staggering 2240 percent over the last 35 years.[12]

Large agribusiness corporations, backed by various institutional investors, have also consolidated control over other segments of the food supply chain, including seeds and fertilisers. This concentration is demonstrated by recent mergers in the global seed and agrochemical sectors, where just four companies—Bayer, Corteva, ChemChina-Syngenta, and BASF—control 60–80 percent of the market share. This concentration of power reduces competition and can lead to higher input prices for farmers.[13]

The financialisation of the system has also altered the landscape of agricultural finance. While it has provided some farmers with better access to capital through credit markets and investment, it has also exposed them to higher levels of debt and financial risk.

Financial entities have also increased their investment in agricultural land, often referred to as "land grabs."[14]

This trend can displace local farmers and communities, disrupt traditional farming practices, and lead to monoculture farming, which, in turn, may lead to adverse environmental impacts.

With the growing emphasis on maximising shareholder value, this trend can often lead to strategies focused on financial metrics and short-term gains rather than long-term investment. It can also lead to cost-cutting measures affecting product quality, labour conditions, and sustainability practices. Unfortunately, a small number of large transnational companies—predominantly based in high-income countries but with complex tax domiciles—concentrate power and profit across several of our food supply systems.[15]

Similarly, the increasingly centralised global food and farming system often affects emerging countries with low food self-sufficiency, rising import dependency, and greater reliance on foreign aid. At the same time, emerging countries face more significant food security risks due to their weak financial systems and inadequate agricultural financial regulatory systems compared to developed countries.[16]

Regrettably, our current commercial food system's emphasis on maximising profit and shareholder value does not bear responsibility for the high costs of its resultant activities from a societal, health, or environmental perspective. "When the institutions of money rule the world, it is perhaps inevitable that the interests of money will take precedence over the interests of people", David Korten.[17]

With future population and urbanisation growth, as well as increasing environmental, health and societal costs, there is no doubt that our current food system is no longer sustainable in its current form.

A team of MIT graduate students, led by two young scholars, Dennis and Donella Meadows, first published their book, "The Limits to Growth," in 1972, during the counterculture period.[18] This slim book, peppered with computer-generated graphs and written in a clear, dispassionate language, delivered a seemingly extreme argument at the time: If 1970 rates of economic growth, resource use, and pollution continued unchanged, then modern civilisation would face environmental and economic collapse sometime in the mid-twenty-first century.

The book sold 12 million copies, was translated into thirty-seven languages, and remains the top-selling environmental title ever published. In many ways, it helped launch modern environmental computer modelling and began our current globally focused ecological debate. The book was first published over 50 years ago, and according to their predictions, we have just another 25 years to go.

The FAO's (Food and Agriculture Organisation) most recent report on food and agriculture, published in 2023, estimates that the global quantified hidden costs of agrifood systems reached approximately $ 12.7 trillion in 2020.[19] This includes environmental hidden costs from GHG and nitrogen emissions, water use, and land-use change; health hidden costs from productivity losses due to unhealthy dietary patterns; and social hidden costs from poverty and productivity losses associated with undernourishment.

14.2 OUR ANIMAL-SOURCED MEAT DILEMMA

The human diet has a long history of consuming animal-sourced foods, dating back at least 2-3 million years. Indeed, as discussed earlier, according to the 'expensive tissue hypothesis,' our increased brain size and cognitive capacity were made possible by the supply of energy and nutrients from animal-sourced meat and other edible animal-derived items, including those beyond traditional red meat and poultry.

Today, this preference is driven by several other factors, including biological, psychological, cultural, and historical elements. From a biological perspective, animal products provide about 36 percent of the total food supply's calorie content. They supply our bodies' physical requirements for vitamin A, retinol (74 percent bioavailable), vitamin B-12 (65 percent bioavailable), folate (67 percent), niacin or nicotinamide (67 percent), pantothenic acid (80 percent), riboflavin (61 percent), thiamin (82 percent), and vitamin B-6 (83 percent).[20]

They are also a rich source of iron, a vital mineral for human health. Iron is used by our bodies to make proteins, such as haemoglobin in red blood cells, which carry oxygen from the lungs to the rest of the body, and myoglobin, a protein found in muscle cells. Iron also produces hormones and enzymes for energy production and helps build a fully functioning immune system.

Just 100 g of cooked beef will provide an entire day's recommended protein, vitamin B-12, and zinc intake and substantially contribute to meeting the riboflavin and iron recommendations.[21] Animal products also provide flavour and taste, containing compounds such as glutamate, which enhances their umami flavour. Other animal-derived products, such as eggs and milk, also have a high biological value since they provide high amounts of protein and calcium. 100 g of milk can provide substantial amounts of calcium, vitamin B-12, vitamin A, and riboflavin.[21]

From a psychological perspective, many people are attracted to meat's texture, aroma, and flavour, which can create a satisfying and pleasurable sensory experience. One's familiarity with meat or preference is often developed during upbringing, and both biological needs and psychological desires can influence cravings for meat.

From a cultural perspective, many cultures have traditional dishes centred on meat. Eating meat is also a cultural practice passed down through generations and is often associated with celebrations, major calendar events, and social gatherings.

Animal by-products are also used extensively in the food processing industry. For example, gelatin is a protein derived from the processing of animal bones, cartilage, and skin. It is widely used in various food sectors as a gelling agent, stabiliser, clarifier, emulsifier, or a thickening agent in canned foods and sausages. A foaming agent used to improve the texture and reduce the melting speed of ice cream, or in the beer industry as a clarifier.

Gelatin can also be produced from fish scales or skins. These are high in collagen and can be used to make gelatin with different melting and gelling temperatures compared to mammalian gelatin.[22]

An interesting debate in food is whether fish counts as meat or not. The classification of fish as a meat source has long been a subject of discussion. On the one hand, meat refers to skeletal muscle, fat, tissues, and innards, which, by definition, would mean there is no question that fish is not meat. On the other hand, as defined by certain religions, fish being cold-blooded as opposed to mammalian animals was how meat was described in the 9th century. Religions such as Catholicism, Lutheranism, and some Orthodox churches do not classify fish as meat.[23]

Outside of the food industry, the history of candle and soap making also goes back to Mesopotamian times when fat rendered from slaughtered animals was combined with lye, created from wood ash, to produce the first soaps. It was in 1837 that William Procter, an English candle maker, and James Gamble, an Irish soap maker, merged their businesses in Ohio to form P&G. Today, the same company is one of the world's leading FMCG brands, offering a range of products including laundry detergents, dishwashing liquids, and other soap products.[24]

In some societies, consuming meat can be a symbol of wealth and status. Historically, meat was a luxury item, and its consumption was a sign of affluence. Indeed, wealth and meat consumption remain strongly correlated globally. Western nations average more than 100 kgs (220 pounds) of meat per person annually, while the poorest African nations average less than 10 kgs (22 pounds).[25] Global demand for meat is growing: over the past 50 years, since the introduction of industrial animal agriculture, meat production has more than quadrupled. The world now produces more than 350 million tonnes each year. Growing populations demand higher livestock products—meat, milk, and eggs—to satisfy an increasing appetite for animal-based foods.

Likewise, as people become wealthier in developing nations, they tend to consume more animal products.[26]

China and Brazil have seen meat production increase approximately 10-fold from 7 and 3 million tons per annum in 1970 to 93 and 30 million tons, respectively. China's average meat consumption per head of population now equates to 62.75 kg per annum, up from just 3.35 kg in 1961.[27]

Globally, pork is the most widely consumed meat commodity per capita. This is followed by poultry, beef/buffalo meat, and mutton/goat, with other types of meat and fish making up a smaller fraction of the overall consumption.

Consumption trends vary significantly across the world. In China, pig meat accounts for around two-thirds of per capita meat consumption. In Brazil, beef and buffalo meat dominates, accounting for over half of consumption. New Zealanders have a much stronger preference for mutton & goat meat, while Portugal's meat preference is for fish and seafood.

While the global meat industry provides food and livelihood for tens of millions of people who depend on livestock farming, it also becomes a significant sustainability challenge arising from the need to balance a growing global demand for meat with the world's finite resources available for its production.

Terrestrial animal agriculture, including meat and dairy production, already accounts for more than three-quarters of agricultural land use. Over 90 percent of farmed animals globally are living in factory farms, with ramifications as detailed earlier. This includes an estimated 74 percent of farmed land animals (vertebrates only) and virtually all farmed fish.[28]

Animal agriculture is also responsible for between 11 and 20 percent of global greenhouse gas emissions and more than 30 percent of global methane (CH_4) emissions.[29]

Methane production by ruminant animals, such as cows and sheep, is linked to global warming.[30] Methane's lifetime in the atmosphere is significantly shorter than that of carbon dioxide (CO_2), but CH_4 is more efficient at trapping radiation than CO_2. Pound for pound, the comparative impact of CH_4 is 28 times greater than that of CO_2 over a 100-year period.[31]

A projected rise in food-related greenhouse gas emissions could seriously impede efforts to limit global warming to acceptable levels. Global temperatures are projected to warm by approximately 1.5 degrees Celsius (2.7 degrees Fahrenheit) by 2050 and 2–4 degrees Celsius (3.6–7.2 degrees Fahrenheit) by 2100. The best science we have indicates that to mitigate the worst impacts of global warming, we must achieve net-zero carbon emissions globally by no later than 2050.

Animal agriculture is also a significant contributor to deforestation and land-use change, necessitating the expansion of pastureland and increased animal feed production. Monogastric animals, including pigs and poultry, require relatively high-quality dietary protein. Soybean meal is the preferred source of protein in pig diets, and nearly 80 percent of the world's soy production is now used as animal feed. Soy production has more than doubled over the last two decades.[31]

Soy is often grown on land cleared through deforestation, converting forests and grasslands into mono-crop soy production. Brazil is the world's largest producer of Soy at 156 million metric tons, and as per the FAO, the rate of deforestation in Brazil was 1.70 Mha (17 000 sq. kilometres) per year between 2015 and 2020.[32]

Of the approximately 1200 million tons of corn produced globally, roughly 60 percent is now used directly as animal feed. The United States is the world's largest producer of corn. In 2023, US farmers planted approximately 94 million acres of significantly mono-cropped corn to produce approximately 347 million tons of maize.[33] The harvest is used domestically for animal feed, ethanol production, and export. Brazil has also become a significant exporter in recent years, often second only to the US in export volumes.

Indirectly, as animal feed products are also a key component of corn and wheat milling economics, actual grain tonnage used for animal feed statistics is much higher than just whole grain numbers.

If a corn or maize miller makes flaking grits for cornflake production, he would be lucky to achieve a finished product extraction rate of just 55 percent. Similarly, he would expect an extraction rate of approximately 60–65 percent for brewing grits, used extensively as an adjunct for beer production. The balance of corn milled is then sold as animal feed at a discounted price relative to the cost of the whole grain.

Similarly, with wheat milling, the global production volume of wheat reached almost 785 million metric tons in the marketing year of 2023/24.[34] A wheat miller would expect a 70-75 percent extraction rate, depending on the wheat quality and flour produced. The balance, often referred to as "wheat offal," is then sold to animal feed producers or feedlots at a discounted price relative to whole grain wheat prices.

From a future food meat supply perspective and considering the rising demand for animal-sourced meat products, if global trends continue, the amount of grain feed required by 2050 is expected to double, with estimates ranging from 1.8 to 2.3 times current usage.[35]

To put this in context, we would need to produce even more crops to raise livestock than to sustain ourselves.

Research suggests that if everyone adopted a plant-based diet, we could reduce global land use for agriculture by 75 percent. This significant reduction would be possible thanks to reduced land used for grazing and a minor need for land to grow crops. The research also shows that cutting out beef and dairy (and substituting with chicken, eggs, fish, or plant-based foods) has a significantly more significant impact than eliminating chicken or fish.[36]

While animal-sourced foods have long been a vital part of the human diet, the debate surrounding sustainability concerns has also led to a clash between those who advocate for abolishing livestock farming and those who defend it as a critical element of healthy and sustainable societies.[37]

Livestock land use is extensive because it requires approximately 100 times more land to produce a kilocalorie of beef or lamb compared to plant-based alternatives, such as peas or tofu.[38] However, livestock animals have the potential to be fed without competing with direct human nutrition, and in doing so, can play a vital role in resource-efficient, regenerative, and agroecological food systems. The type of land used to raise cows or sheep differs from that used for crops such as cereals, potatoes, or beans. Livestock can be raised on pasture grasslands or on steep hills where it is impossible to grow crops.

Advocates of plant-based diets have gained popularity in recent times due to their potential to reduce the environmental impact of our food choices. While the number of vegetarians and vegans remains relatively low compared to those who eat meat, reports from the Intergovernmental Panel on Climate Change (IPCC) and the EAT-Lancet Commission on Food, Planet, Health recommend that individuals who consume high amounts of meat consider adopting a flexitarian diet to help create a more sustainable food system.[38] Similar to the Mediterranean diet, the flexitarian diet primarily consists of vegetables, fruits, whole grains, and unsaturated fats. It does include some high-quality meat, dairy, and sugar, but in much smaller quantities than are typically consumed in more affluent societies. By reducing their meat consumption, flexitarians can contribute to a more sustainable food system and help alleviate the strain on resources such as water, land, and energy.

However, with a large proportion of the global population dependent on animal products, and even more who would benefit from increased consumption, as well as many agricultural workers dependent on incomes derived from their production, demonising meat consumption is not the way forward. If we utilise farm animals for what they excel at—converting by-products from the food system and grass resources into valuable food and manure—they can significantly contribute to a healthy human food supply.

Balancing the rising trends of animal-sourced meat and grain crop feeding will require a multifaceted approach that combines sustainable practices, technological innovation, policy support, and global cooperation.[39] As will be seen in the next chapter, addressing these issues holistically may create a more sustainable and equitable food system.

14.3 OUR NOMENCLATURE: *HOMO SAPIENS*, AND OBESITY

In the 18th century, Carl Linnaeus developed a classification system to identify every species he knew using a standard nomenclature: a genus name followed by a species name. His system, expressed in Latin, the scientific language of the time, was based on similarities in obvious physical traits. Each species was given a unique two-word Latin name. His two noteworthy books, Species Plantarum[40] and System Naturae, were published in 1753 and 1758, respectively. The Linnean system remains in use today and is widely accepted as the foundation of our current botanical and species nomenclature system.

Linnaeus' most famous scientific name is probably the name he gave humans, *Homo sapiens*. The Latin noun homō (genitive hominis) refers to a "human being" or "man" in the generic sense of humanity. Sapiens is derived from a Latin word that means 'wise' or 'astute'. Indeed, over centuries, the meaning of the word *Homo sapiens* has since changed to mean modern humans, as defined by the Oxford Dictionary, "the kind or species of human that exists now".

Never a modest man, Linnaeus, at age 33, claimed he was the scientific equal to Newton and Galileo. He wrote that he was the greatest botanist and zoologist, having revolutionised an entire science. He was concerned exclusively with similarities in bodily structure. Without the DNA evidence we have today, he faced only the problem of distinguishing *Homo sapiens* from other primates grouped in the great ape category (gorillas, chimpanzees, orangutans, and gibbons). He concluded that apes differ from humans in numerous bodily and cognitive features. The latter element one supposes inspired his term for *Homo sapiens*.

Names for other human species were randomly introduced in the second half of the 19th century as new fossil findings were discovered. *Homo neanderthalensis* (an extinct species of human being) 1864, *Homo erectus* (erect) 1892, and *Homo hablis* (the handyman) 1959–60. The name *Homo* was retained as the genus name for these findings, and the species name was adapted as a more descriptive name, in Latin, of their attributes.

But let's suppose, for a moment, an alien landed on Earth today and was given the same task that Carl Linnaeus completed in 1758. Suppose he was tasked with defining a descriptive Latin name for the species of humans that now inhabit the earth—all 8 billion *Homo sapiens*.

From a biological perspective, the same alien might concur with the *Homo* genus nomenclature because of our biological characteristics, genetic makeup, and evolutionary history. From a behavioural and cultural perspective, humans can also be described as tool-makers, complex communicators, and highly social beings. Similarly, from a technological standpoint, we have developed advanced technology and the ability to alter natural habitats globally.

However, if the alien judged us from an impact perspective on the only habitat we have he/she would surely conclude that humanity seems intent on destroying its life support system. Humanity's ecological footprint now exceeds the Earth's biocapacity—the planet's ability to regenerate natural resources—by 70 percent.[41]

Wildlife populations are in freefall around the world, driven by human overconsumption, population growth and intensive agriculture. Tanya Steele, chief executive at WWF, says: "We are wiping wildlife from the face of the planet, burning our forests, polluting and over-fishing our seas and destroying wild areas. We are wrecking our world – the one place we call home – risking our health, security and survival here on Earth."[42] Hardly the attributes of a "wise" or "astute" being despite the impressive cognitive capability.

So, supposing the alien could not agree with our cognitive abilities, sapiens, as a suitable descriptive species differentiator, what other physical attribute could they use to distinguish us from our biological classification as primates?

Despite sharing 98.6 percent of the same DNA as our primate classification, the main differentiator today would have to be that we have now become "the fat primate." Or, to give it a suitable Latin name, *Homo adipem*.

Herman Pontzer, Professor of Evolutionary Anthropology at Duke University, describes how the great apes—gorillas, chimpanzees, bonobos, and orangutans—live relatively inactive lives.[43] They typically spend eight to 10 hours a day resting, eating, and grooming, then sleep for nine to 10 hours at night. Chimpanzees walk only about 4 km a day, gorillas less. However, the great apes are remarkably healthy despite their low physical activity levels. Their blood pressure does not increase with age, diabetes is rare, and although chimpanzees have high cholesterol levels, their arteries do not harden and block. And great apes mostly do not grow fat.

On the other hand, as humans, as per the World Health Organisation, in 2022, 2.5 billion adults (18 years and older) are now overweight. Of these, 890 million are living with obesity. Over 390 million children and adolescents aged 5–19 years are also overweight, including 160 million now living with obesity.[44]

The figures for children are even more startling when one considers the fact that just 2 percent of children and adolescents aged 5–9 were obese in 1990 (31 million young people); by 2022, that figure has increased over fivefold to 160 million young people.

While obesity was once only considered a problem in high-income countries, a propensity for being overweight is also on the rise in low- and middle-income countries. In Africa, the number of overweight children under five has increased by nearly 23 percent since 2000. Almost half of the children under five who were overweight or living with obesity in 2022 live in Asia.[44]

In many low- and middle-income countries, there is a double burden of malnutrition. This means that while children in these countries are more vulnerable to inadequate prenatal, infant, and young child nutrition, they are also exposed to high-fat, high-sugar, high-salt, energy-dense, and nutrient-poor foods. These foods are often cheaper and lower in nutritional value. As a result, undernutrition and obesity traits can be found in the same developing countries.[45]

Childhood and adolescent obesity can have significant consequences for both physical and mental health. It leads to an increased risk and earlier onset of non-communicable severe diseases (NCDs) such as type 2 diabetes, cancers and cardiovascular problems. It can have adverse effects on mental well-being, school performance, and overall quality of life, often compounded by societal stigma, discrimination, and bullying. Likewise, children who struggle with obesity are also more likely to continue to grapple with weight issues into adulthood, placing them at a higher risk for developing non-communicable diseases in later life.[46]

As previously discussed, while genetic and hormonal factors may play a role in contributing to obesity, excessive consumption of calories from unhealthy foods and drinks is undoubtedly a primary factor in childhood obesity. Encouraging the early development of healthy eating habits is essential during childhood, a time of rapid growth and development, when general eating habits and patterns are formed in the first few years of life.

A summary of international young child feeding recommendations from the EU, Dietary Guidelines for Americans, WHO, UK, Nordic Nutrition, and Canada all recommend breastfeeding for the first six months of life, where possible. From six months of age, breast milk should be supplemented with adequate, safe, and nutrient-dense foods.[47]

Numerous studies have also shown that the more variety of tastes, textures, colours, and mouth feels a baby is exposed to, the more likely those children are to accept new foods later on. Between 12 and 48 months, dietary fibre intake should increase from 10 to 20 g per day. Sugar-sweetened beverages (such as soda, fruit drinks, sports drinks, and energy drinks) are not necessary for a child's or adolescent's diet and are not recommended. Natural foods prepared with little or no added sugar should be the primary choice for children and adolescents.

However, as we have also seen, the food manufacturing industry, driven by profit, market share, and growth opportunities, does not necessarily follow recommended dietary advice. The global baby food market supply business reached over $88 billion in 2022 and is expected to grow at a CAGR of more than 6.3 percent from 2023 to 2032.[48]

So, how do these processed baby food products compare to Mom's breast milk?

While the flavour of breast milk varies, it contains water, fat, carbohydrates (lactose), proteins, vitamins, minerals, and amino acids, depending on the mother's diet. The lactose—about 7 percent of breast milk has about 200 different sugar molecules, which serve a wide range of purposes. As newborns do not have the means to digest these sugars, researchers believe they play a vital role in feeding the newborn baby's microbiome gut bacteria. This helps develop the infant's immune system and establishes a healthy balance of bacteria in their gut.[49]

From a food processing perspective, given that most pureed fruit, vegetable, and grain products are cooked at high temperatures in a retort system, consider a large commercial pressure cooker or a similar heat treatment plant. How do they retain the natural flavours and nutrients of the raw ingredients? The short answer is that they can't, particularly with extended-shelf-life products.

Think of Fresh milk pasteurised at 71 degrees C for 15 seconds. It tastes good but needs refrigeration and only has a shelf life of 5–10 days. If the same milk is processed at 130–150 °C, it becomes UHT (ultra high temperature). The shelf life is extended to 6 months, and it no longer requires refrigeration, but the taste profile is off. If the same product is evaporated, canned, and cooked using a retort system, its shelf life is increased to 6–18 months; however, the product no longer retains its natural milk taste.

While preserving food by way of heat treatment and glass jars goes back to Nicolas Appert's time and his quest to provide Napoleon with extended shelf life for army rations. It was a similar quest by the United States Army, Natick R&D Command, working with Continental Flexible Packaging and the Reynolds metal company that first invented the retort pouch in the mid-seventies.[50]

It was around 2007 when Plum Organics and Ella's Kitchen first started packaging baby food in pouches, and since then, several manufacturers have offered ready-to-use squeeze pouches globally. In baby food aisles in supermarkets today, at least 25 percent of that space is devoted to a wide array of different pouch products and flavours.

The reasons behind the rapidly increased market share of squeeze pouch baby foods are not hard to discern; as any parent will tell you, they provide the ultimate convenience for feeding babies and toddlers. Pouches require no prep, no refrigeration, no mess, and virtually no cleanup. They are ideal for 'time-poor' parents seeking what they perceive as a healthy food option for their children. Food pouches marketed for infants and young children nearly always contain an integral spout with a valve that means the contents can be directly sucked from the bag itself. This provides a mess-free and self-feeding option for parents and infants, eliminating the need for bowls or washing up.

The problem is the processing and high sugar content, which make the product attractive to infants and help mask the taste of processing. Likewise, the child's development of oral motor skills, including tongue lateralisation, chewing, gagging, and swallowing, required for eating regular food, is hindered.

Despite recommended guidelines as detailed earlier, some of these products are marketed as suitable for infants from four months old. The "from four-month-old" options were also found to have a higher sugar content (8.7 ± 3.6 g/100 g) than toddler (12-month) equivalents. While many products state "no added sugar" on the front of the pack, this statement is misleading. Natural sugars in pureed fruit and vegetable products are classified as free sugars and can range up to 17.5 (g/100 g) in the same pouches.

A recent paper published in BMC in 2023 from Australia tested 276 pouch products from 15 manufacturers. It concluded that squeeze pouch products were nutritionally poor, high in sugars, not fortified with iron, and posed a clear risk of harm to the health of infants and young children if fed regularly. In the USA, a Baby Food Facts Report found that most infant squeeze pouches do not support recommendations for promoting healthy eating habits, and the marketing of these pouches is misleading about the actual nutritional content of the products and their sweetness levels.[51]

Dried cereal products for toddlers, which are often supplemented with milk or water, do not fare much better. A European Union report included 4196 infant foods and 502 processed cereal-based foods. The report showed that 1359 (31.9 percent) of baby foods had added or free sugars, and 1167 (27.4 percent) had one or more types of sugar among the top five ingredients. According to the same report, the average energy of dry commercial baby food complementary cereals was 69.1 g of carbohydrates with 15.3 g of total sugar.[52]

And so begins a child's induction into the world of (ultra) processed foods and high sugar content. It sets a food taste preference for both adolescents and adults for life.

In the mid-20th century, ultra-processed foods (UPFs) were almost non-existent. However, the significance of a global transition to UPFs in the food supply chain has become alarming. In Mexico and Chile, nearly a third of the total calories consumed by the population come from UPF. This number rises to almost 60 percent in the UK and USA. The highest consumption of UPF is observed in children and adolescents, with UPF contributing to over 50 percent of their daily total calorie intake in several countries.

Conversely, Italy had the lowest levels (about 10 percent), and the latter being inversely associated with adherence to a Mediterranean diet. While there is still controversy surrounding the use of the term "ultra-processed food" (UPF), it is undeniable that the global rise in obesity has coincided with the increase in UPF consumption worldwide. Multiple studies have confirmed that higher UPF consumption is linked to weight gain. Evidence suggests that individuals consuming high amounts of UPF may have up to a 50 percent higher risk of developing obesity compared to those consuming less UPF.

Looking beyond a suggested fictional name change for our species, *Homo sapiens*, from a broader economic perspective. As per a UK government report, Obesity now costs the NHS (their national health system) around £6.5 billion a year and is the second most significant preventable cause of cancer. Over one in four (26 percent) adults and 23.4 percent of children aged 10–11 years in England are living with obesity.[53]

According to a recent Senate report, the US government will spend approximately $283 billion on obesity-related direct health costs in 2023, rising to $526.5 billion by 2033. As a result, the total projected government expenditure on obesity-related direct health costs over the 2024–2033 10-year budget window is $4.1 trillion.[54]

According to the BMJ (British Medical Journal), More than half (51 per cent) of the global population will be living overweight or obese within 12 years unless prevention, treatment, and support improve. The World Obesity Federation (WOF) has said in its World Obesity Atlas 2023 that the economic impact of overweight and obesity on the world is set to reach $4.32tn—nearly 3 percent of global gross domestic product—annually by 2035.[55]

Johanna Ralston, the CEO of the World Obesity Federation, said, "Let's be clear: the economic impact of obesity is not the fault of individuals living with the disease." It results from high-level failures to provide the environmental, healthcare, food, and support systems we all need to live happy, healthy lives."

REFERENCES

1. B. Marr, *Business trends in practice: the 25+ trends that are redefining organizations*, Wiley, Hoboken, NJ, 2022.
2. M. Friedman, *Capitalism and freedom*, University Of Chicago Press, Chicago, 1982.
3. H. Takahashi, *Everything Originated From Milk: Case Study Of Nestle*, World Scientific, 2021.
4. Nestle, https://www.nestle.com/investors/shares-adrs/dividends.

5. ShareAction, https://shareaction.org/news/shareholders-file-health-resolution-at-nestl%C3%A9 March 2024.
6. I. Quinn, Campaign group ShareAction defiant after losing Nestlé HFSS vote, The Grocer, 19th April 2024.
7. Farm Europe, Are futures the future for farmers? https://www.farm-europe.eu/travaux/are-futures-the-future-for-farmers-2 April 2016.
8. CME Group, https://www.cmegroup.com/markets/agriculture/grains/corn.html.
9. F. Kaufman, https://foreignpolicy.com/2011/04/27/how-goldman-sachs-created-the-food-crisis/.
10. Statista, Number of futures and options contracts traded worldwide from 2013 to 2022 (in billions), https://www.statista.com/statistics/377025/global-futures-and-options-volume/.
11. A. J. Mead, *The Complete Financial History of Berkshire Hathaway*, Harriman House Limited, 2021.
12. C. Katje, If You Invested $1000 In Coca-Cola Stock When Warren Buffett Did, Here's how much You'd Have Now, Benzinga.com, 2023.
13. Collectif, *Concentration in Seed Markets*, OECD, 2018.
14. M. Fairbairn, *Fields of Gold*, Cornell University Press, 2020.
15. V. Nair, How is the world tackling tax avoidance by multinational companies? Economics Review, 2023, https://ifs.org.uk/articles/how-world-tackling-tax-avoidance-multinational-companies.
16. FAO, *State of Food and Agriculture, 2021 (SOFA) : making agrifood systems more resilient to shocks and… stresses*, S.L., Food & Agriculture Org, 2021.
17. D. C. Korten, *When corporations rule the world*, Berrett-Koehler, San Francisco, 1995.
18. D. H. Meadows, D. L. Meadows, J. Randers and W. W. Behrens, III, *The limits to growth: a report for the Club of Rome's project on the predicament of mankind*, Universe Books, New York, 1979.
19. FAO, *Hidden costs of agrifood systems at the global level*, The State of Food and Agriculture, 2023
20. S. M. S. Chungchunlam and P. J. Moughan, Comparative bioavailability of vitamins in human foods sourced from animals and plants, *Crit. Rev. Food Sci. Nutr.*, 2003, **64**(31), 11590–11625.

21. S. P. Murphy and L. H. Allen, Nutritional Importance of Animal Source Foods, *J. Nutr.*, 2003, **133**(11 Suppl 2), 3932S–3935S.
22. S. K. Kim, *Seafood Processing By-Products: Trends and Applications*, Springer New York, New York, NY, 2014.
23. R. S. Ellwood, *The Encyclopedia of World Religions*, Infobase Publishing, 2008.
24. D. Dyer, F. Dalzell and R. Olegario, *Rising tide: lessons from 165 years of brand building at Procter & Gamble*, Harvard Business School Press, Boston, Mass., 2004.
25. BBC Which countries eat the most meat? https://www.bbc.co.uk/news/health-47057341.
26. W. Winders and E. Ransom, *Global meat: social and environmental consequences of the expanding meat industry*, The Mit Press, Cambridge, Massachusetts, 2019.
27. H. Ritchie, P. Rosado and M. Roser, *Meat and dairy production*, Our World in Data, 2023.
28. K. Anthis, J. Reese Anthis, *Global Farmed & Factory Farmed Animals Estimates*, Sentience Institute, 2019.
29. P. K. Malik, *Livestock production and climate change*, Cabi, Wallingford, Oxfordshire, UK, 2015.
30. United States Environmental Protection Agency, https://www.epa.gov/ghgemissions/overview-greenhouse-gases#methane.
31. H. El-Shemy, *Soybean and Health*, 2011.
32. H. S. Klein and F. V. Luna, *Feeding the World*, Cambridge University Press, 2018.
33. The National Agricultural Statistics Service (NASS), Agricultural Statistics Board, United States Department of Agriculture (USDA), June 30, 2023.
34. M. Shahbandeh, *Wheat - statistics & facts*, Statista, 2024.
35. R. Bhat, *Future foods: global trends, opportunities, and sustainability challenges*, Academic Press, London, 2022.
36. H. Ritchie, *If the world adopted a plant-based diet, we would reduce global agricultural land use from 4 to 1 billion hectares*, https://ourworldindata.org/land-use-diets.
37. H. Steinfeld, *Livestock in a Changing Landscape*, 2010, vol. 1.
38. K. Hirvonen, Y. Bai, D. Headey and W. A. Masters, Affordability of the EAT–Lancet reference diet: a global

analysis, The Lancet Global Health, 2019, https://www.thelancet.com/journals/langlo/article/PIIS2214-109X(19)30447-4/fulltext.
39. T. Beal, C. D. Gardner, M. Herrero, L. L. Iannotti, L. Merbold, S. Nordhagen and A. Mottet, Friend or Foe? The Role of Animal-Source Foods in Healthy and Environmentally Sustainable Diets, *J. Nutr.*, 2023, **153**(2), 409–425.
40. C. von Linné, *Species plantarum*, 1764.
41. D. Thorpe, *The "One Planet" Life*, Routledge, 2014.
42. BBC, Wildlife in 'catastrophic decline' due to human destruction, scientists warn, https://www.bbc.co.uk/news/science-environment-54091048#:~:text=Wildlife%20is%20%22in%20freefall%22%20as,and%20time%20is%20running%20out.%22.
43. H. Pontzer, *BURN: new research blows the lid off how we really burn calories, stay healthy, and lose weight*, Avery Pub Group, S.L., 2022.
44. WHO, Obesity and overweight, https://www.who.int/news-room/fact-sheets/detail/obesity-and-overweight.
45. L. R. Marquez, *Improving health care in low—and middle-income countries: a case book*, Springer International Publishing, Cham, 2020.
46. W. Burniat, T. J. Cole, I. Lissau and E. Poskitt, *Child and Adolescent Obesity*, Cambridge University Press, 2006.
47. gov.uk., SACN., 'Feeding young children aged 1 to 5 years' Annex 1 SACN 2023.
48. Precedence Research, Baby Food Market Size to Attain USD 155.32 Bn by 2032, https://www.precedenceresearch.com/press-release/baby-food-market.
49. S. H. Rubin, The ABCs of Breastfeeding, AMACOM Div American Mgmt Assn, 2008.
50. A. Ghani and M. M. Farid, *Sterilisation of food in retort pouches*, Springer, New York, London, 2011.
51. K. A. Brunacci, L. Salmon, J. McCann, K. Gribble and C. A. K. Fleming, *The big squeeze: a product content and labelling analysis of ready-to-use complementary infant food pouches in Australia*, BMC Public Health, 2023.
52. M. A. Theurich, *et al.*, Nutritional Adequacy of Commercial Complementary Cereals in Germany, *Nutrients*, 2020, **12**(6), 1590.

53. DHSC Media Team Obesity, public health, Government plans to tackle obesity in England, https://healthmedia.blog.gov.uk/2023/06/07/government-plans-to-tackle-obesity-in-england/.
54. US Government information, H.Res. 1047, https://www.congress.gov/118/bills/hres1047/BILLS-118hres1047ih.pdf.
55. World Obesity, Economic impact of overweight and obesity to surpass $4 trillion by 2035, https://www.worldobesityday.org/resources/entry/world-obesity-atlas-2023.

CHAPTER 15

Towards a More Sustainable Future Food System

15.1 COSTING THE CHANGE REQUIRED

Johan Rockström is an internationally recognised Swedish scientist who works on global sustainability issues. He co-founded the Stockholm Resilience Centre and is currently the director of the Potsdam Institute for Climate Impact Research in Germany. He also co-chaired the EAT-Lancet Commission on Healthy Diets from Sustainable Food Systems, published in 2019,[1] and he is co-chair of Future Earth and the Earth Commission. He summarises the significance of our food system in a single sentence:

"The global food system holds the future of humanity on Earth in its hand."

With projected future population and urbanisation growth, as well as considerable associated environmental and health costs that are not reflected in the price of the food we produce, there is no doubt that our current food system and its economics need to change.

How can we create a sustainable food system for the future?

Sustainability is a slightly amorphous word. Do we want to *sustain* a status quo that has catalysed multiple global challenges, including climate change, biodiversity loss, and widespread food inadequacies such as inadequate nutrient intake, micronutrient deficiencies, malnutrition and diet-related chronic disease?

Food and Us: The incredible story of how food shapes humanity
By Seamus Higgins
© Seamus Higgins 2025
Published by the Royal Society of Chemistry, www.rsc.org

Sustainable development has a much more explicit definition. It focuses on the idea that progress and growth should contribute to positive environmental, economic, and social outcomes while minimising or avoiding negative impacts.[2]

Given that food production is a fundamental necessity of life, how do we reconcile humanity's need for a stable food supply and subsequent health and wellness with a commercial food system operating in an economic environment that demands continual profit growth? A fixation on narrowly defined efficiency, productivity, and perpetual growth has resulted in a discipline blinkered to our current ecological cataclysm.

As our introductory question asked about increasing medical and scientific proficiency, helping us multiply, could that same resource of sociotechnical innovation and immense human ingenuity—our sapiens, cognitive ability!—assist us in achieving the change in direction our current food system requires?

Or, put more succinctly, how could that same resource assist us in developing a more sustainable future food system from an environmental, economic, social and regulatory framework perspective?

Addressing these challenges is not straightforward as they are interlinked and create a web of complexity that will require multifaceted solutions. A change in one aspect of the food system can have ripple effects throughout. For example, adopting more sustainable farming practices may impact productivity, food prices, and availability with repercussions on nutrition and indirect land-use change.

Regarding environmental sustainability, Solutions must balance the immediate need for food with the planet's long-term health, requiring a nuanced understanding of ecosystems and resource management.

From an economic and social perspective, future policies must also consider the economic implications for farmers, consumers, and businesses, as well as other social factors such as population growth, cultural dietary preferences, and food accessibility. Daunting as the challenges of transforming food systems may be, there are reasons to be hopeful.

Despite human population tripling from 2.5 billion in 1950 to 8 billion people in 2024 and an emphasis on sustainability, discussions focused on efficiently feeding an exponentially growing population, this challenge may diminish as birth rates start to decline across the globe.

In keeping with Heraclitus' concept of ever-present change, a 2020 study published in The Lancet projects that the global population will peak in 2064 and decline to 8.8 billion by 2100.[3] The world's overall fertility rates are dropping, with women having one child fewer on average than they did around 1990.

In more than half of all countries and areas, the average number of live births per woman is below 2.1—the level required for a population to maintain a constant size. Half of all countries in the world today have fertility rates below the population replacement level. Even low- and middle-income countries, such as India and Mexico, associated with rapid population growth, are now experiencing slowing rates. High-income Asian countries, including Japan, Hong Kong, Singapore, South Korea, and Taiwan, are experiencing ultra-low fertility rates.

One of the most significant factors shaping world population growth is urbanisation. Over the past century, the largest migration in human history has occurred and continues today as people move from rural areas to cities. As we have already explored, in 1900, about 14–16 percent of the world's population lived in urban areas. Today, almost 60 percent of the global population lives in urban areas, which rises to 80 percent in developed countries. The transition from rural to city life changes the economic incentives and drawbacks of having large families. In rural areas, many children mean additional help with work, but in urban areas, many children mean more mouths to feed.

However, as we have seen with increasing incomes and urbanisation, diets change—people consume more animal-sourced foods, sugar, fats and oils, refined grains, and processed foods. Ignoring the consequences of today's food systems locks the world onto a course that escalates these adverse effects disastrously. Recent research from various institutions has come to similar conclusions. These reports conclude that the hidden costs of our current food systems are mortgaging our future and undermining future productive potential by well over 10–15 trillion US dollars a year.

A recent 2024 report by the Food System Economics Commission (FSEC), an independent commission established to assess comprehensive options for transforming the food system, summarises the findings of a four-year investigation based on rigorous economic modelling, in-depth literature reviews, and case studies.[4]

These include Health costs, which FSEC estimates to be at least US$ 11 trillion. They measure the economic costs of ill health due to food systems through their adverse effects on labour productivity. These are driven by the prevalence of non-communicable diseases, including diabetes, hypertension, and cancer, which can be attributed to food. A large share of this burden is borne by people living with obesity. If current trends continue, the global adoption of diets high in fats, sugar, salt, and ultra-processed foods will increase the number of obese people worldwide by 70 percent to an estimated 1.5 billion in 2050, or 15 percent of the expected global population. The direct medical costs of treating the health consequences of overweight and obesity have already been estimated by others to rise from 600 billion USD today to almost 3 trillion by 2030.

Environmental costs are estimated at $3 trillion per year and reflect the negative impacts of today's food systems on ecosystems and climate, including the effects of current agricultural land use and food production practices.[5]

According to UNEP's 2018–2019 Frontiers report, nitrogen pollution costs the global economy between US$340 billion and US$3.4 trillion annually, taking into account its impact on human health and ecosystems.

Finally, food systems contribute to structural poverty through high food costs and the low incomes of many workers in the food production sector. The incidence of poverty tends to be higher in agriculture than in other segments of the food system (Figure 15.1).

But what if the costs of transforming the food system were remarkably modest compared to the expected benefits?

The FSEC estimates a cost range of between 200 and 500 billion USD annually, depending on the extent to which the expenses of ensuring food affordability for the most vulnerable are factored in. Estimates of these benefits, measured as reductions in the unaccounted costs of food systems amount to at least $ 5 trillion per year. When the full effects of a global food system

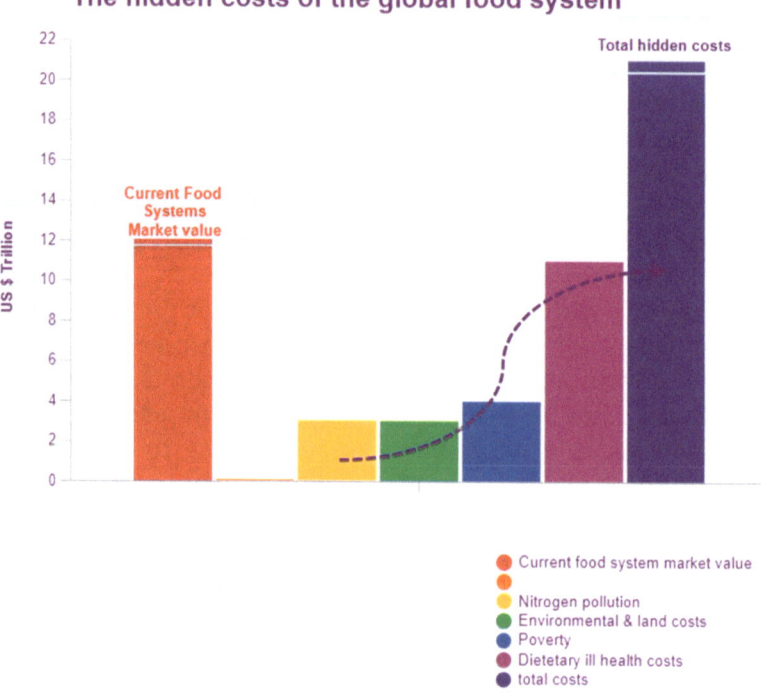

Figure 15.1 The hidden costs of the global food system. Data from ref. 4.

transformation on incomes are factored in, estimates of its benefits could rise to 10 trillion US$ annually.

Leaving economics aside, for now, picture the benefits.

- Undernutrition could be eliminated by 2050, and cumulatively, 174 million lives saved from premature death due to diet-related chronic disease, compared to current trends.
- A shift to environmentally sustainable production in agriculture reverses biodiversity loss, reduces demand for irrigation water and almost halves nitrogen surplus from agriculture and natural land.
- The food system could become a net carbon sink by 2040. As part of a more extensive sustainability transformation, which includes the energy sector. Helping to ensure that global warming is limited to below the 1.5 °C Paris Climate target.

This alternative future would unfold differently in various parts of the world. A shift to healthier diets would entail notably higher consumption of fruits, vegetables, and nuts in South and Southeast Asia, as well as legumes in China. Meanwhile, consumption of animal-sourced food would have to decrease drastically in high—and middle-income regions.

As we have seen in various chapters outlined in this book, humanity's history is a story of ingenuity in the face of multiple challenges. We already have several success stories to inspire us.

More recently, in 1987, the world came together to protect the world's ozone layer by signing the Montreal Protocol on Substances that deplete the Ozone Layer.[6] This agreement aimed to reduce and eventually eliminate the use of manufactured chemicals, specifically chlorofluorocarbons (CFCs), which were depleting the ozone layer. The protocol was enacted in 1989 and has since been amended several times to increase ambition and reduce targets. Since its adoption, every country on Earth has signed the Montreal Protocol—the only treaty to have been universally ratified. It is now widely considered a triumph of international environmental cooperation.

A trio of scientists—Paul Crutzen, Frank Rowland, and Mario Molina—who would later win the Nobel Prize in Chemistry, proposed that human emissions of chlorine substances were depleting ozone in the stratosphere, which in turn protects life on Earth from the sun's UV (Ultra-Violet) radiation.

They faced strong resistance and denial from industrial and political players. At the time, the chairman of DuPont—the largest global manufacturer of CFCs—said the theory was: "a science fiction tale … a load of rubbish … utter nonsense". Indeed, leading CFC producers formed the "Alliance for Responsible CFC" to coordinate their efforts and launched intense PR campaigns discrediting the ozone depletion theory. Anne Gorsuch, the US's head of the Environmental Protection Agency, dismissed ozone depletion as an environmental scare. However, the visual imagery of a growing ozone hole was hard to ignore, and it finally pressured governmental and industrial actors to take action.

The Montreal Protocol has been highly successful, with countries eliminating nearly 99 percent of banned ozone-depleting substances. Since then, there has been a notable recovery of the ozone layer in the upper stratosphere and decreased exposure to harmful ultraviolet rays from the sun. The Antarctic ozone hole is slowly healing, although the atmosphere will not fully recover until after 2070, as CFCs have atmospheric lifetimes of 50 years or more.

The world, in turn, found better ways to gas our fridges and air conditioners, invented less harmful aerosols, and used fiscal measures in rich countries and development aid for the poorer ones to help make the transition possible.

DuPont closed its last CFC-11 and CFC-12 production facilities in September 1995. It remains a successful chemical company, with a market capitalisation of 33.52 billion as of 2024.[7] Indeed, its food ingredient company, DuPont Danisco, remains one of the world's largest producers of food emulsifiers, including guar gum and carrageenan. Its mission statement has also subsequently changed to read, "DuPont puts science to work by creating sustainable solutions essential to a better, safer, healthier life for people everywhere."[8]

One wonders if the world could adopt a similar mission statement to address our current food systems dilemma.

15.2 BREAKING THE "JUNK" FOOD CYCLE

Michael Jacobson highlighted the term "junk food" as part of the counterculture movement in the early 1970s.[9] Armed with a PhD in microbiology from MIT, he and two other scientists established the Centre for Science in the Public Interest in the USA in 1971. By the 1990s, CSPI's flagship publication, the Nutrition Action newsletter, had become the country's most prominent food and nutrition-focused publication.

Before the concept of ultra-processed foods emerged, junk food was defined as food containing high levels of refined sugar, white flour, polyunsaturated fats, salt, and various food additives, but lacking protein, vitamins, and fibre. Fifty years later, junk food (UPF) has become a global phenomenon. Diet and nutrition have emerged as significant health risk factors worldwide, even in poorer countries. In the Western world, the consumption of fresh, whole foods has declined, while highly processed foods have become increasingly common.

Junk food is often cheaper than whole foods, widely available, and heavily marketed. Processed meats, such as bacon, deli meats, and hot dogs, as well as fast foods like burgers, pizzas, and French fries, are high in calories but low in nutrients. Sugary drinks, such as sports drinks, energy drinks, and soft drinks, are junk food. Snacks include potato chips, crackers, and sweets or candy.

Other foods that could be considered junk food include white bread, ready-to-eat breakfast cereals, bakery products, ice cream, and frozen yoghurt. Recent research (July 2024) from the Universities of Cambridge and Bristol has found that UK adolescents now consume around two-thirds of their daily calories from ultra-processed foods (UPFs) or junk food.

Adolescents from disadvantaged backgrounds consume a higher proportion of their calorie intake from UPFs than adolescents from less underprivileged backgrounds (68.4 percent compared with 63.8 percent).[10]

Since 1970, the UK has seen a 200 percent increase in soft drink purchases in grams/person/week, similar to takeaway ready meals and sweetened breakfast cereals. Crisp consumption has risen by over 4000 percent.[11]

The junk food cycle is a theory that explains the rise in food-related health issues in recent decades. It can be described as a reinforcing feedback loop characterised by the industry and unhealthy eating habits that lead to weight gain and obesity, heart disease, cancer, and a weakened immune system.

As discussed in previous chapters, one of the main reasons people prefer junk food is its taste. Junk food is designed to be enjoyable, often combining high levels of sugar, salt, and fat to create an irresistible flavour. Poor diets have primarily developed due to our evolved appetites and natural preference for high-calorie foods. When we eat junk food, the reward circuits in our brains respond to food cues, activate and release the neurotransmitter dopamine, impeding self-control and satiety and motivating us to eat more.

When these consumer characteristics are combined with the economic incentives of a food system that prioritises volume, turnover, and profit, food production companies capitalise on satisfying this demand. The industry's competitive nature prioritises using a higher volume of low-cost calories and utilising inexpensive ingredients. Similarly, it will focus on added processing and ingredients to enhance taste and palatability, aiming to achieve the so-called "bliss point."

Processed food often undergoes numerous refining steps that strip away essential nutrients, leaving behind a product that is not only lacking in healthful components but also teeming with artificial additives and preservatives. The industry also increases its efforts to reduce perishability, extend shelf life, and increase production runs to create a better economy of scale.

Of course, increasing sales also encourages a significant investment in advertising to achieve better sales volume, growth, and profitability. Food, beverage, and restaurant companies spend almost $14 billion annually on food advertisements in the United States. More than 80 percent of this food advertising promotes fast food, sugary drinks, candy, and unhealthy snacks.[12]

Globally, food and drink stalwarts such as Coca-Cola spend up to 10 percent of sales revenue on marketing, averaging $4 billion annually.[13] In 2023, Nestlé's marketing and advertising expenses were 7.7 percent of sales, an increase of 80 basis points from the previous year, equating to just over 7 billion CHF ($ 8.296 billion).[14]

Social media has also revolutionised how food brands engage with their audience. The rising trend of social media, influencer marketing and digital advertising has also seen significant growth, with food being the second most active industry in influencer marketing.

However, social media is a two-way street. While Internet advertising attempts to influence our food choices across targeted demographics and persuade us to buy certain foods, consumers using the same media also have the power to create the opposite effect. As Kellogg's CEO Gary Pilnick recently discovered when faced with a massive backlash following his comments in a 2024 CNBC interview about a new Kellogg's ad campaign. The campaign encouraged people to give "chicken the night off" and add cereal to the weekly dinner line-up, including Kellogg's Frosted Flakes, Froot Loops, and Frosted Mini-Wheats. He also mentioned that families with strained finances could cope by eating "cereal for dinner".

People interpreted Pilnick's comments as financial advice for low-income families and criticised the campaign for promoting unhealthy eating habits for children. A Guardian newspaper article, "Let Them Eat Flakes", and several others highlighted the online criticism aimed at Gary Pilnick and Kellogg's after the interview. One TikTok user sarcastically mentioned Pilnick's annual base salary of $1 million and more than $4 million in incentives and questioned whether he would feed his kids Kellogg's cereal for dinner. Another user reacted by saying: "What kind of dystopian hellscape is this? Give the peasants' cereal for dinner!"[15]

Due to the nature of digitally served advertising and the limited availability of data on reach and engagement, it is challenging to accurately assess the scale of junk food marketing on these channels. However, food and drink companies are now investing millions in online marketing to ensure their products remain at the forefront of our minds, and they would not do so if advertising did not work. As per the Junk Food Cycle diagram below, investment in advertising and marketing plays a crucial role in promoting the same reinforcing feedback loop (Figure 15.2).

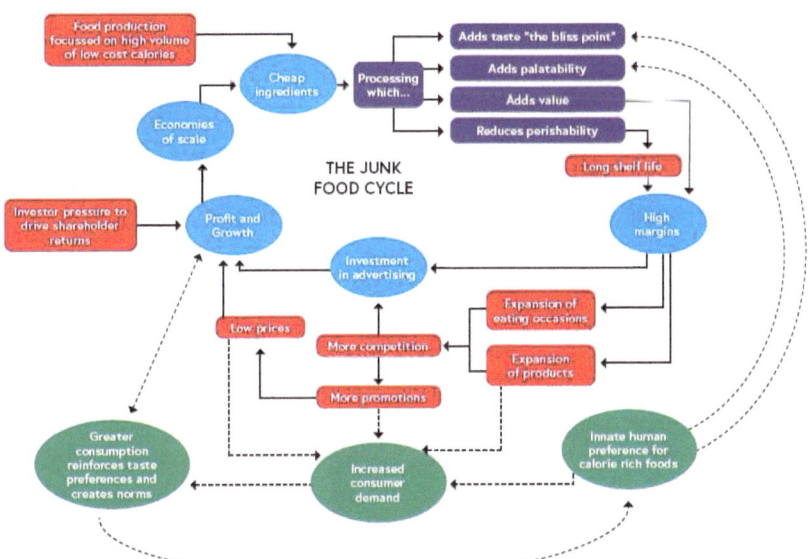

Figure 15.2 The Junk Food Cycle.[16] Reproduced from ref. 16 with permission from Henry Dimbleby.

The same cycle becomes even more amplified when it transfers from manufacturers to retailers. With intense competitive price pressure in the consumer retail market, this sector further perpetuates the junk food cycle by oversupplying, promoting, or discounting high-fat, high-sugar, or high-salt (HFSS) foods.

As retailers focus on driving revenues through volume promotions and impulse purchases, promotions on unhealthier foods tend to be highly expandable. Junk food promotions are more likely to increase turnover and incremental consumer spending in the retail sector.

Finally, from a consumer's perspective, our innate preference for calorie-rich foods reinforces taste preferences, increases consumption and consumer demand, and reinforces the entire junk food cycle.

Efforts to break this cycle of junk food consumption will necessitate a comprehensive approach that engages individuals, families, communities, food manufacturers, and policymakers. The prevailing notion that improved education and awareness of the harmful effects of junk food and the advantages of a balanced diet will enable individuals to take charge of their health and escape the cycle is not a practical solution.

As Henry Dimbleby points out in "The Plan," the UK's independent national food strategy document. Of the 689 UK diet-related government policies launched between 1992 and 2020, just under half (43 percent) put the onus on individuals to change their behaviour, and 37 percent were policies that supported healthier eating but still required individuals to make better food choices.[16] As per current UK overweight and obesity statistics, it's evident that these programmes, which require individuals to change their behaviour, have not been effective.

He also notes that the premise behind these programs is based on the assumption that individuals make well-informed, rational choices about their food intake and possess the motivation, resources, and capability to do so. This assumption is seriously flawed for several reasons, including the lack of financial resources, time, kitchen skills, facilities, or the necessary motivation. As the numerous examples in this book show, people do not and sometimes cannot make rational choices about their food intake.

Regarding sustainability and health, big food companies can also be victims of their own success. Mass appeal is great for the bottom line but leaves these companies in a tight spot. On the one hand, they have to do a ton of public relations to show off their efforts in sustainability and health, but on the other hand, they cannot point out that their core products are one of the reasons why we have a junk food cycle in the first place.

Industry "cash cow" products that create significant negative externalities but are deeply ingrained in consumers' eating and drinking habits will continue to thrive as long as people continue to buy them.

While local and national governments also play essential roles in providing healthier food and ensuring food security for their populations, to my knowledge, no country has implemented a comprehensive range of updated, evidence-informed strategies to promote a healthier and more equitable food system.

From a food marketing perspective, one particularly intriguing aspect of food preferences is the divergence and convergence of food choices across different generations. What if we considered a future endeavour to break the junk food cycle, taking into account these norms and generational preferences?[17]

Baby boomers (born 1946–1964): Enjoy classic meals, global flavours, and simplicity. They are known for their inclinations towards traditional values and prefer familiar comfort foods often tied to nostalgia. They appreciate home-cooked conventional meals and are more likely to value recognisable ingredients, their favourite brands, and, given their age, products that support healthy ageing.

Generation X (born 1965–1980) is marked by independence and adaptability. They tend to seek a balance between convenience and health, favouring foods that offer both nutrition and convenience. Starbucks has Generation X to thank for bringing the brand to the forefront with its embrace of coffeehouse culture in the 1990s.

Millennials (born 1981–1996) are often described as adventurous, health-conscious, and open to trying new cuisines and flavours. They prioritise authenticity, sustainability, and social responsibility in their food choices. Convenience, coupled with health benefits, heavily influences their decision-making. Millennials were the initial drivers of the plant-based movement, making Beyond Meat a popular choice for Millennials due to its sustainability focus and meat alternative offerings.

Generation Z (born 1997–2012): These digital natives are highly diverse and globally connected. They value food experiences and seek adventurous, Instagram-worthy meals. Health consciousness and ethical considerations heavily influence their food preferences, favouring plant-based options and sustainability. TikTok and Facebook have significantly influenced Gen Z, as they rely on social media platforms like TikTok for food trends and recommendations, which in turn influence their food choices.

Baby boomers, who make up 1.1 billion or 15 percent of the world's population, are more set in their ways when it comes to food choices. Generation X, 23 percent of the global population, or almost 2 billion people, are known for being versatile, resilient and sceptical. They are also more educated than older generations and considered the wealthiest generation in history, responsible for 27 percent of global spending. Now, consider that same generation as future grandparents and a sceptical resource as food advisors to their Millennial and Generation Z children.

While Millennials comprise approximately 23 percent of the global population, Gen. Z has become the world's largest generation, with over 32 percent of the worldwide population. By 2026, they will also make up the majority of first-time parents.

As shown above, Gen Z is more likely to incorporate sustainability and social responsibility into their food choices. Health consciousness and ethical considerations heavily influence their food preferences. As such, as future parents, Gen Z could be expected to have a different approach to parenting than previous generations. So now would be the ideal time to engage with this critical demographic and endeavour to create a new food paradigm for them and their children.

As mentioned in the previous chapter, the world's scientific community agrees that sugar-sweetened foods and beverages are unnecessary in a child's or adolescent's diet and are not recommended. Dietary fibre, freely available in whole and natural foods, should increase from grams per day by 2–5 years to 15 g, 6–11 years, 20 g, 12–16 years, 25 g, and 17+ years, 30 g. We should also look to rebalance our diets by prioritising plants, fruits and vegetables and moderating our intake of animal products.

As enshrined in the United Nations Convention on the Rights of the Child, all children have the right to the best possible health, nutritious food, and education.[18] What if all food system stakeholders, including individuals, families, communities, food manufacturers, and policymakers, could collaborate to develop a new approach that encompasses all elements involved in feeding a population to support a new generation of children?

For now, let's assign a name to them and refer to them as Generation A or Alpha.

15.3 SUPPORTING A FOOD SYSTEM FOR A NEW GENERATION: GENERATION A (ALPHA)

Starting with baby's first foods, given the global scientific agreement on the ideal diet for complementary baby and toddler food, policymakers and governments need to regulate this market to ensure these accepted guidelines are adhered to. This would not only help develop future food tastes for children but also create a level playing field for food manufacturers. It would also help to alleviate food manufacturers' concerns, such as Nestlé's, that "A proportional target (of selling HFSS foods) would require us to weaken valuable parts of our portfolio and create opportunities for competitors without yielding public health benefits."[19]

As will be seen in the next section, the days when sugar was used to mask the taste of food processing or extend shelf life are no longer necessary. Technology has advanced to enable safe food cold processing and treatment for extended shelf life. Likewise, the level and type of sugars produced during processing, which depend on the ingredients and the level of hydrolysis (*i.e.*, time and temperature) used, may no longer be applicable.

Recent research involving parents and their infants aged 6–24 months was conducted where one group tested a high-sugar refined cereal (21 g/100 g), and the other a low-sugar cereal (<1 g/100 g) with 50 percent of whole grain, which represented a 95.2 percent decrease in sugar content. It was found that there were no significant differences between the two groups in terms of infants' overall acceptability (infants' reactions, estimated intake, and relative intake).[20]

Indeed, incorporating whole-grain cereals and utilising more ancient grains could be a promising direction for future children's cereals. Unlike modern grains, many aspects of ancient grains have been rediscovered, including their nutritional and health benefits, as well as their physicochemical properties, which can be utilised in food applications. Ancient grains include varieties of wheat (spelt, Khorasan wheat or Kamut, Einkorn and Emmer); green wheat, barley; wild rice, oats; sorghum; millets, and pseudo cereals of teff, amaranth; buckwheat and quinoa. They often contain more protein, dietary fibre, bioactive compounds, and antioxidant activity, showing improved health benefits. Moreover, as a future parent, ancient grains would also align with Gen Z's values of sustainability and ethical consumption values. These grains typically require fewer agricultural inputs, such as pesticides and fertilisers, and are often grown in smaller, more sustainable farms.

When it comes to liquids, all living things need water to survive. Water is vital to our health. It plays a crucial role in many of our body's functions, including transporting nutrients to cells, eliminating waste, protecting joints and organs, and regulating body temperature. Along with milk, we know plain water is the best drink choice for kids. Children under five should only have milk or water as a fluid source. Not only are these the only tooth-friendly options, but water is also a healthy and inexpensive choice, with zero calories and no added sugar, which can help kids maintain a healthy weight into adulthood.

Public health recommendations also suggest limiting children's consumption of fruit juice in favour of whole fruit due to the juice's high sugar content, lack of fibre from whole fruit, and potential for excessive intake. The next step would be to ensure that this new "Generation A" maintains healthy eating and drinking habits when they grow beyond the five-year-old stage. Their millennial grandparents' taste for foods high in fat, salt, and sugar, as well as the provision of treats high in sugar or fat, would also have to be curtailed.

The Global Action Plan for the Prevention and Control of Non-Communicable Diseases, proposed by the World Health Organization, suggests utilising economic tools (*e.g.*, taxes and subsidies) to discourage the consumption of less healthy options and promote healthy dietary choices.[21]

Over the past few decades, more than 50 countries and jurisdictions have implemented health taxes on sugar-sweetened beverages (SSBs), while 18 countries have taxed foods high in fat, salt, and sugar (HFSS). Excise taxes on ad-quantum are the most common health taxes applied to food and alcoholic beverages. The differentiation of value-added tax (VAT) rates on foods and drinks has also been considered an alternative option for incentivising healthy diets.

In Mexico, an 8 percent tax on non-essential food items with a high-calorie content relative to their weight led to a 6 percent decrease in purchases. In Hungary, a tax on unhealthy foods led to a sustained decline in the consumption of those foods by most consumers.

Henry Dimbleby's "The Plan" proposal for the UK recommended that the government introduce a £3 per kg tax on sugar and a £6 per kg tax on salt sold for use in processed foods, restaurants, and catering businesses. It is also recommended that some of the revenues generated be used to provide or subsidise fresh fruit and vegetables to low-income families.[16]

Regardless of generational categorisation, food-insecure individuals tend to consume fewer fruits, vegetables, and dairy products and have lower intakes of vitamins and minerals. This can lead to unhealthy dietary patterns, where people replace higher-quality foods with more energy-dense foods. Taxes can help to address health inequalities, as less affluent socio-economic groups tend to respond more to price increases following taxation and often derive more significant health benefits from reducing their consumption of taxed products.[16]

Likewise, the tax revenues generated can fund or augment more innovative health food programmes for low-income families, such as a UK scheme, which mirrors similar schemes in the United States. Doctors in deprived London boroughs have begun prescribing fruit and vegetable vouchers to patients. The vouchers are worth between £6 and £8 per week, or £2 per household member. After just eight months of receiving vouchers for fruit and vegetables, 80 percent of participants were eating five portions each day, compared to just 28 percent at the start of the programme.[22]

The health impacts of this change are striking: nine in ten participants have experienced improvements in their physical health, including healthy weight loss, increased energy levels, and enhanced digestion. Over half of the participants reported that their mental health has also improved, as they worry less about financial concerns, including food. With support from social prescribers and healthcare professionals, these positive changes have collectively resulted in a 40 percent reduction in GP (doctor) visits.

Whether rich or poor, better primary and secondary school education could also play a crucial role in shaping the future "Gen. A" children; no country, except Norway, currently has a comprehensive curriculum that fully addresses food literacy.[23] Even though food literacy from primary school to adolescence has the potential to positively influence health, cooking, and sustainability behaviours that can be tracked into adulthood, it could also be instrumental in generating population shifts toward improvements in the food system.

While food literacy as a teaching subject is still emergent and differs from country to country, most definitions are based on a nutrition and food skills component. Curricula focus on cooking and health topics but less on social-cultural, equity, and sustainability issues.

Expanding these concepts more broadly, a definition of food literacy could also incorporate the ability to access, process, and enjoy food in a manner influenced by our complex food system. Tracy Cullen et al. proposed a definition of food literacy that encompasses a more comprehensive, positive relationship built through social, cultural, and environmental experiences with food, enabling people to make informed decisions that support their health.[24]

The ability of an individual to understand food in a way that they develop a positive relationship with it, including food skills and practices across the lifespan in order to navigate, engage, and participate within a complex food system. It is the ability to make decisions to support the achievement of personal health and a sustainable food system considering environmental, social, economic, cultural, and political components.

In addition to food literacy, school meals have also been shown to support children's health and education. In OECD countries which can afford it, free school meals are the most straightforward and least intrusive way to ensure that all children have at least one well-balanced, healthy, and nutritious meal daily. School meal schemes can also drive more environmentally friendly food production. Given their large purchasing quantities, school feeding programmes can also drive purchasing towards more local agro-ecological output.

According to a 2022 report from UNESCO, over nine in ten countries have implemented school health and nutrition (SHN) programmes worldwide.[25] Policies and programmes vary between countries, reflecting priorities, available resources, and capacity differences. Strong country leadership and investment are required to ensure that all children and adolescents are in school and that SHN programmes reach those most in need in the poorest countries and marginalised households. Undernourishment, micronutrient deficiencies and over-nutrition are still significant public health challenges worldwide.

The report emphasises that a holistic approach to caring for learners' health and well-being is one of the most transformative and cost-effective ways to enhance future educational and health outcomes. However, realising this potential will necessitate a shift in perception regarding the role of schools. It should go beyond solely focusing on academic achievement and prioritise the health and well-being of learners as a crucial component of future education.

As future parents, Generation Z, who have never known a time without a device in their hands, has a different relationship with technology than previous generations. This will be even more evident with their children and the next generation, Gen. A., as they grow up in an urbanised, intelligent, and interconnected world of super-fast digital communication.

Generation A consumers' future perspectives will change as advancements in computing power, scientific research, and innovation reshape the current food model. This change will hopefully lead to a more sustainable, environmentally friendly, and health-conscious approach to diet driven by consumer knowledge.

Personal intelligent technologies, such as personalised avatars, will likely monitor dietary preferences and consumption habits.[26] These technologies could draw from a cloud of scientific knowledge to inform and guide future food and beverage purchasing decisions. Thanks to instant access to information from anywhere across the globe, Generation A will be better equipped than any other generation to tackle the problems we cannot solve today.

Subsequently, food manufacturers and their brands will need to respect the environment that these people inherit and do so convincingly: social media and other digital networks will leave nowhere to hide.

With better food education, Gen Alpha may be less concerned with 'more' and instead use their spending power to attain lifestyles with more positive social and environmental benefits.

Around the world, we already see people beginning to question the actual costs of their purchasing choices—wanting to know more about where their food comes from, its environmental impact, and examining the health implications of their diet. Current aspirations of bigger, faster, and 'more' would hopefully shift towards a new type of brighter, cleaner, healthier lifestyle.

15.4 SCALING UP MORE SUSTAINABLE TECHNOLOGY AND INNOVATION

Yuval Noah Harari, the author of Sapiens: A Brief History of Humankind, believes that humans are the most successful species on Earth because of their ability to cooperate flexibly and in large numbers.[27] Scientists attribute our success to technological innovation, group collaboration, and communication. In addition to these abilities, humans excel in imagination, ingenuity, and inventiveness, which have also been crucial to our success.[28]

Consider shared fiction: Humans can create shared fiction that can become literal truths. For example, we use money to exchange value because we trust that others believe in it. Our imagination and language skills also enable us to create and share fictional stories. Developing technological innovation based on these attributes has allowed humans to reach the moon, split the atom, and decipher DNA.

What if we could use these same attributes to change the direction of our current food system for the better? As mentioned in earlier chapters, approximately ten thousand years ago, the first domestication of plants and animals occurred, marking the beginning of humans' breeding of plants and animals for food and labour. These could be considered wild macroorganisms, ranging from cows and sheep to wheat and barley.

Today's rapid advances in new technology allow us to manipulate microorganisms far more than our ancestors could have imagined. We can now isolate microorganisms entirely from macroorganisms and harness them directly as superior and more efficient units of nutrient production.[29]

What if we could engineer alternative products or processes to augment our current and future animal-sourced meat dilemma?

Cultivated meat, also known as cultured meat, is genuine animal meat (including seafood and organ meats) produced by directly cultivating animal cells. This production method eliminates the need to raise and farm animals for food. Cultivated meat is composed of the same cell types that can be arranged in the same or similar structure as animal tissues, thereby replicating the sensory and nutritional profiles of conventional meat.

Dutch scientist Mark Post unveiled the first cultivated meat burger on live television in 2013. Two years later, the first four cultivated meat companies were founded.[30] As of late 2022, the industry had grown to more than 150 companies on six continents, backed by $2.6 billion in investments, each aiming to produce cultivated meat products. More than a dozen companies have also been established to develop technology solutions throughout the value chain.

If these products are produced using renewable energy, they could reduce greenhouse gas emissions by up to 92 percent and land use by up to 90 percent compared to conventional beef. Additionally, commercial production is expected to occur entirely without antibiotics and will likely result in fewer incidences of foodborne illnesses due to the lack of exposure risk from enteric pathogens.

Regarding the use of precision fermentation to change protein sources, what if we could also utilise similar microorganism technology to produce food directly from the air we breathe?

As explored earlier and as Dr Lisa Dyson explains in her TED talk, "A Forgotten Space Age Technology Could Change How We Grow Food," microorganisms are already an integral part of our everyday lives. If you enjoy a glass of wine or a beer after a long workday, then you are enjoying a product of microbes. If you enjoy cheese, bread or yoghurt, you also enjoy a microbe-enabled product.[31]

Now, consider a single-cell organism or microbe, similar to yeast, that utilizes hydrogen, which can be sustainably produced through electrolysis from H_2O (water), and uses carbon dioxide as a feedstock instead of emitting it. Hydrogenotrophic methanogens play a vital role in the carbon cycle and are found in numerous oxygen-free environments. They are free-living cells in freshwater and marine environments, cold sediments, and hydrothermal vents.

The unique attribute of these single-cell organisms is that they grow in hours instead of months. They reproduce asexually, utilising hydrogen and carbon dioxide as their energy source. This means we could produce an additional sustainable food source much quicker than we do today. They grow in the dark, allowing them to thrive in any season, geography, and location. They can grow continuously in vertical containers that require minimal space. They require just 1 litre of water per kg of protein produced, which is 112 000 times less than beef, almost 8000 times less than legumes, and nearly 6000 times less than soy.[32]

The harnessing of these same microbes to create this unique food source is no longer science fiction! In addition to developments with Dr Lisa Dyson's Air Protein Company in California, Solar Foods of Finland has taken this process beyond bench testing. Their first commercial factory was commissioned in the spring of 2024.[33] They're making a product called Solein from the same microbes. The product has all the essentials of a perfect food source: sixty-five to seventy percent protein, roughly ten to fifteen percent dietary fibres, five to eight percent fats, and three to five percent mineral nutrients. It also contains all nine essential amino acids.

As a new functional food, Solein easily enhances various sweet and savoury foods, making it ideal for virtually any food imaginable. It is highly nutritious, vegan, and caters to every diet. The macronutrient composition of the cells is very similar to that of dried soy or algae, but it is more versatile since it has a pleasant umami flavour and mild aroma.

Just as we look to produce future food from air and single-cell microbes, we can also use similar technology to create environmentally friendly fertilisers. Building on research from the Kyoto

University of Japan and its spin-off company, Symbiobe, they are producing a bacterial product called Symbiobe that could dramatically reduce greenhouse emissions and avoid other environmental problems caused by the overuse of fertilisers.[34]

Fertilisers supply three principal nutrients to plants: nitrogen, phosphorus and potassium. The nitrogen included in most commercial fertilisers comes from ammonia, which is manufactured at an enormous scale by combining nitrogen gas from the air with hydrogen gas at high temperatures and pressures in the energy-intensive Haber–Bosch process.

When the Haber–Bosch process was invented more than a century ago, it revolutionised farming, producing the world's first synthetic fertilisers to produce sufficient food for the Earth's growing population.

The trade-off was that the same process comes with significant environmental consequences. Ammonia production accounts for up to 2 percent of the world's energy output. It releases approximately half a billion tonnes of carbon dioxide annually, accounting for roughly 1.8 percent of global carbon dioxide emissions.[35]

The contribution of inorganic synthetic fertilisers to greenhouse gas emissions doesn't stop after manufacture. When inorganic fertiliser is applied to a field, microbes in the soil convert some of its nitrogen to nitrous oxide (N_2O), a gas with a very high greenhouse warming potential. N_2O gases directly emitted from agricultural soils to the atmosphere are a major contributor to global warming, as N_2O has a global warming potential 273 times greater than carbon dioxide (CO_2). As synthetic fertilisers deliver nitrogen to plants almost instantaneously, this rapid action also exacerbates nitrous oxide formation and nitrogen runoff. Excessive application leads to increased nitrates in waterways, which in turn promote algae growth and waterway eutrophication. However, just as food from the air can enhance existing food systems, Symbiobe or similar air fertilisers could be blended with other organic fertilisers, such as manure or compost, to achieve a more sustainable alternative to inorganic mineral nitrogen fertilisers.

Another promising technology currently being developed worldwide is plasma technology, which can generate a nitrogen alternative to the Haber–Bosch process. Making nitrogen-based fertiliser in a plasma reactor involves a process known as nitrogen fixation. While 78 percent of the air surrounding us consists of nitrogen, the gas does not react with other elements (it is chemically inert). This makes it hard for plants to use.

Nitrogen fixation solves this problem. It converts the nitrogen (N_2) from the air into NO_x, which reacts with oxygen and water to form nitrate (NO_3). This can then be used as an ingredient for liquid fertiliser.

Utilising state-of-the-art plasma technology to produce affordable fertilisers for small farmers. It may sound like magic, but it has now become reality. Researchers at Eindhoven University of Technology (TU/e) have developed a small plasma-powered plant that produces a nitrogen-based liquid fertiliser using only sunlight, water, and air. "The plant is easy to set up, sustainable and very efficient", says TU/e researcher Fausto Gallucci, who, together with partners in Africa, Germany and Portugal, has successfully tested the device in Uganda. "We now aim to bring the mini-plant to market, making it available to farmers worldwide."[36]

When developing new technology for food preservation systems, the possibilities for cold plasma technology, which creates a fourth state of matter, become endless. Cold plasma is a non-thermal technology that involves making a partially ionised gas at low temperatures by applying an electric field to the gas. Consider the sun, stars, or lightning as natural examples of plasma, a state of matter characterised by electrically conductive charged particles.

In Australia, inspired by traditional Aboriginal lore of grain farmers burying their rainstick or *"chuggura"* covered in metal oxide to attract lightning to make their seeds grow faster, Rainstick is a new Australian start-up working on bioelectrically plasma enhanced seeds for bigger, faster, and more sustainable crop yields.[37]

Electrically produced cold plasma can also be utilised as a novel non-thermal food processing technology, employing energetic, reactive gases to inactivate contaminating microbes on meats, poultry, fruits, and vegetables. It becomes a fast, cost-effective, and eco-friendly methodology for eliminating harmful microbes without affecting the food's nutritional or sensory qualities. Cold plasma can also improve food colour, flavour, and texture, enhancing the functional properties of proteins and altering their allergenicity. Consider its potential as an alternative to retorting baby food at high temperatures and pressures to extend shelf life.

The quest to develop other novel non-thermal processing technologies has become an essential part of food industry development, as a response to increasing shelf life and food safety without adversely affecting the quality of natural foods. Thus, providing products with higher functionalities and reduced processing and environmental costs. These novel processing technologies, such as high hydrostatic pressure (HHP), pulsed electric fields (PEF), ultrasonics (US), Cold Plasma (CP), and pulsed UV (ultra violet) light, have recently gained attraction by numerous researchers and new company start-ups as microbial inhibition and natural food preservation techniques.

While the phrase "let thy food be thy medicine, and thy medicine be thy food" is often attributed to Hippocrates, there is no concrete evidence that he articulated these words! Nonetheless, the idea resonates with the core principles of Hippocratic philosophy, which is to "do no harm" or at least "to help".[38] More recently, we have seen a renewed focus on the relationship between food and medication.

The discovery of antibiotics, such as penicillin, in the 1930s is still considered one of the most significant advances in medical history. Before insulin was first used as a treatment for diabetes in 1922, children with type 1 diabetes were expected to live only around 1.5 years after their diagnosis. In adults, only 1 in 5 would be alive 10 years after their diagnosis.[39]

Today, statins have become one of the most widely prescribed medications globally, with over 200 million people taking them daily to lower cholesterol levels. The tablets are marketed as a medicine to help reduce the risk of heart attacks, strokes, and other cardiovascular diseases, ultimately benefiting heart health.[40]

In December 2017, Novo Nordisk's injectable version of semaglutide, marketed under the Ozempic brand, was approved by the US Food and Drug Administration (FDA) for use in people with type 2 diabetes. The European Medicines Agency (EMA) approved it in February 2018.

Semaglutide belongs to a group of drugs called GLP-1 agonists. These drugs increase the levels of incretins—hormones— naturally produced by the stomach in response to food intake. It works by helping one's body produce more insulin when needed. It also reduces the amount of glucose, or sugar, produced by the liver and slows down how quickly food is digested.[41]

Because the drug slows down the rate of food digestion, it also reduces appetite, causing individuals to eat less. Hence, a more recent iteration as a weight loss drug.

In 2021, the FDA approved Novo Nordisk's Wegovy brand (semaglutide) as an injection (2.4 mg once weekly) for chronic weight management in adults with obesity or overweight conditions and at least one weight-related condition, such as high blood pressure, type 2 diabetes, or high cholesterol.

In the three years since semaglutide was approved for treating obesity, it has taken America by storm. Image-conscious influencers, actors and well-heeled financiers are not the only users. Already, one in eight American adults has been on glp-1 drugs. Novo Nordisk, the maker of semaglutide, branded as Ozempic for diabetes and Wegovy for weight loss, and Eli Lilly, which makes tirzapeptide branded as Mounjaro, another effective alternative, have together added around $ 1 trillion. in market value since 2021.[42]

With over two-fifths of the world overweight or obese, demand for glp-1 drugs is voracious. Pharma companies are racing to make them work as pills, which would be cheaper to produce than jabs and to reduce their side effects. Generic versions for older GLP-1 agonists are also entering the market. Semaglutide will come off patent in Brazil, China, and India in 2026, and eight similar drugs are currently in development in China.

Overweight patients on semaglutide have been found to suffer fewer heart attacks and strokes; the benefits, astonishingly, seem to be largely independent of how much weight is lost. Trials show that glp-1 agonists reduce chronic kidney disease in diabetics, and there are signs they may lessen brain shrinkage and cognitive decline in Alzheimer's. Studies of health records suggest that they may help with addictions, too, as the drugs appear to reduce inflammation and interact with mechanisms linked to cravings and feelings of reward. People already on glp-1 drugs in America were less likely to overdose on opioids or abuse cannabis or alcohol.[43]

It is still early days, and reported side effects are being investigated, but GLP-1 receptor agonists appear to have all the makings of one of the most successful classes of drugs in history. While it could be argued that drugs such as statins or semaglutide treat the effect of our current food system rather than its cause, there is no doubt that as they become cheaper and easier to use, they promise to dramatically improve the lives of billions of people—with profound consequences for industry, the economy, and society.

Recent research (December 2024) from Cornell University, aptly named "The No-Hunger Games: How GLP-1 Medication Adoption is Changing Consumer Food Purchases," highlights the fact that US households with at least one GLP-1 user reduce their food spending by approximately 6 percent within 6 months, with higher-income households cutting their expenditures by nearly 9 percent.[44]

The study highlights the largest spending reductions among users in ultra-processed categories. On average, spending on products such as chips, baked goods, sides, and cookies fell between 6.7 and 11.1 percent. The researchers say, "Our findings highlight the potential for GLP-1 medications to significantly reshape consumer food demand, a trend with increasingly important implications for the food industry as adoption continues to grow."

It's perhaps more striking to see that leading food companies such as ConAgra Brands and Nestlé are actively embracing the market shift by launching new lines of frozen meals branded as "GLP-1 Friendly" and "Vital Pursuit".[45] These offerings are designed to be "portion-aligned" to address their consumers' reduced appetites. They are packed with high levels of protein and are good sources of fibre. They also contain essential nutrients, including vitamin A, potassium, calcium, and iron.

Could this "new" food trend be a positive sign of a significant shift in global food company offerings?

When we consider our current global food system and its complexity, it is clear that food systems are integral to both people's health and the sustainability of our planet. As we have seen over the past 3 million years, our existence depends on both.

Indeed, an observer focused entirely on the challenges our current food systems create might say, "The future is bleak." On the other hand, if one focuses on the opportunities and benefits that social and innovative change could bring, one might say, "The future looks bright."

Perhaps a more realistic observation might be "the future is uncertain".

Tackling the complex challenges of global food systems, ecosystems, and climate change will require transformative, systems-level changes rather than just minor adjustments.

Achieving these goals will depend on the collaboration of all key stakeholders, including leaders from diverse industries, politicians and policymakers, international agencies, research institutes, academia, farmers' associations, NGOs, and other relevant organisations.

The global cost of the recent COVID-19 pandemic is estimated to range between $8.1 trillion and $29.4 trillion, depending on the scope of analysis. A detailed 2020 analysis estimated $16 trillion, including health-related losses. By 2023, mortality costs alone were estimated to be $29.4 trillion globally.[46]

These figures underscore the pandemic's profound economic and societal financial impact worldwide.

As discussed in this final chapter, we possess the necessary resources to achieve the change required for a future food system; financially, the investment required would be less than the cost of the recent COVID-19 pandemic. From a sociotechnical perspective, we also possess the skills, ingenuity, and expertise to create and support a more sustainable food environment that benefits society and the health of our only planet.

From a historical perspective on "Food and Us," we have seen how the food we eat has played an integral role in shaping society, our overall health, and our well-being. From a "Food and Us" future perspective, prioritising health over wealth for our children's future and creating a more sustainable world for generations to come should be considered priceless.

REFERENCES

1. W. Willett, J. Rockström, B. Loken, M. Springmann, T. Lang, S. Vermeulen, T. Garnett, D. Tilman, F. DeClerck, A. Wood and M. Jonell, Food in the Anthropocene: the EAT–Lancet Commission on healthy diets from sustainable food systems, *Lancet*, 2019, **393**(10170), 447–492.
2. J. Agyeman, R. D. Bullard and B. Evans, *Just sustainabilities: development in an unequal world*, Earthscan, London, 2003.
3. S. E. Vollset, *et al.*, Fertility, mortality, migration, and population scenarios for 195 countries and territories from 2017 to 2100: a forecasting analysis for the Global Burden of Disease Study, *Lancet*, 2020, **396**(10258), 1285–1306.
4. Food Systems Economics Commission, Global Policy Report, 2024.
5. World Health Organisation, Obesity and Overweight Key Facts, March 2024.
6. D. Kaniaru, *The Montreal Protocol: Celebrating 20 Years of Environmental Progress: Ozone Layer and Climate Protection*, Cameron May, London, 2007.
7. Yahoo Finance, Valuation measures, DuPont de Nemours, Inc., 6/30/2024.
8. D. Pont, 2023 Sustainability Report, DuPont, 2023.
9. A. F. Smith, *Fast Food and Junk Food: An Encyclopedia of What We Love to Eat [2 volumes]*, ABC-CLIO, 2011.
10. Y. Chavez-Ugalde, *et al.*, Ultra-processed food consumption in UK adolescents: distribution, trends, and sociodemographic correlates using the National Diet and Nutrition Survey 2008/09 to 2018/19, *Eur. J. Nutr.*, 2024, 2709–2723.
11. N. Whittle, The crunch, the flavours, the rituals: how crisps became a British snack obsession, The Guardian Newspaper, 20th Oct. 2024.
12. N. Freudenberg, *At what cost: modern capitalism and the future of health*, Oxford University Press, New York, NY, 2021.
13. M. Ridder, Coca-Cola Company. Coca-Cola Company's advertising expense from 2014 to 2023 (in billion U.S. dollars) [Graph], Statista, https://www.statista.com/statistics/286526/coca-cola-advertising-spending-worldwide/.
14. Nestle, Nestle reports full-year results for 2023. Nestle 22 Feb. 2024.

15. R. A. Vargus, Let them eat Flakes: Kellogg's CEO says poor families should consider 'cereal for dinner', The Guardian Newspaper, 27th Feb. 2024.
16. H. Dimbleby *et al.* National food Strategy, The plan, Recommendations chapter 16 UK Gov. 2021.
17. H. Herring, *Connecting Generations*, Rowman & Littlefield, 2019.
18. United Nations, *The United Nations Convention on the Rights of the Child*, Brill, 1989.
19. I. Quinn, Nestlé tells health activists they are targeting the wrong company, The Grocer, 14th Mar. 2024.
20. L. Sanchez-Siles, S. Román, J. F. Haro-Vicente, M. J. Bernal, M. Klerks, G. Ros and Á. Gil, Less Sugar and More Whole Grains in Infant Cereals: A Sensory Acceptability Experiment With Infants and Their Parents, *Front. Nutr.*, 2022, **9**, 855004.
21. J. S. Yang, Of markets and rights: Discourse in the 2008 and 2013 Global Action Plan for the Prevention and Control of Non-communicable Diseases, *Ann. Global Health*, 2016, **82**(3), 472.
22. E. Wilton, *Press release: Charity urges all parties to commit to national fruit and veg on prescription programmes*, Alexandra Rose Charity, 2024.
23. K. Smith, *et al.*, How Primary School Curriculums in 11 Countries around the World Deliver Food Education and Address Food Literacy: A Policy Analysis, *Int. J. Environ. Res. Public Health*, **19**(4), 2019.
24. T. Cullen, J. Hatch, W. Martin, J. W. Higgins and R. Sheppard, Food Literacy: Definition and Framework for Action, *Canadian J. Dietetic Pract. Res.*, 2015, **76**(3), 140–145.
25. UNESCO, Reimagining Our Futures Together: a new social contract for education, 2022, https://unesdoc.unesco.org/ark:/48223/pf0000379707.locale=en.
26. A. Cesario, *Personalized Medicine Meets Artificial Intelligence*, Springer Nature, 2023.
27. Y. N. Harari, *Sapiens: a Brief History of Humankind*, Harper Perennial, New York, 2011.
28. E. T. Higgins, *Shared reality: what makes us strong and tears us apart*, Oxford University Press, New York, NY, 2019.
29. C. Tubb and T. Seba, *Rethinking Food and Agriculture 2020–2030*, Rethink X, 2019.
30. Evan, D. G. Fraser, D. L. Kaplan, L. Newman and R. Y. Yada, *Cellular Agriculture*, Elsevier, 2023.

31. L. Dyson, A forgotten Space Age technology could change how we grow food, TED@BCG Paris May 2016, https://www.ted.com/talks/lisa_dyson_a_forgotten_space_age_technology_could_change_how_we_grow_food.
32. J. Jolly, Eating light: Finnish startup begins making food 'from air and solar power', The Guardian Newspaper, 19th Apr. 2024.
33. Solar Foods Finland, Solein, https://solarfoods.com/.
34. Kyoto University Symbiobe, A novel fertilizer with a reduced carbon footprint, https://www.nature.com/articles/d42473-024-00097-0.
35. The Royal Society, *Ammonia: zero-carbon fertiliser, fuel and energy store*, Royal Society, UK, 2020.
36. TU/e, Eindhoven University of Technology, Using plasma technology to feed the world, Feb. 2021.
37. A. Brayton, Deep tech and traditional knowledge spark path to _SOUTHSTART CSIRO 28th Feb 2024.
38. J. Anderson, *Let Food Be Thy Medicine and Medicine Be Thy Food Hippocrates*, Independently Published, 2018.
39. M. Bliss and A. Li, *The discovery of insulin*, University Of Toronto Press, Toronto, London, 2021.
40. John Hopkins Medicine, How Statin Drugs Protect the Heart, https://www.hopkinsmedicine.org/health/wellness-and-prevention/how-statin-drugs-protect-the-heart.
41. D. Bagchi and S. Nair, *Nutritional and therapeutic interventions for diabetes and metabolic syndrome*, Elsevier Academic Press, London, Boston, 2012.
42. The Economist: It's not just obesity, Drugs like Ozempic will change the world, Oct. 2024.
43. Science Media Centre, expert reaction to MHRA's approval of semaglutide to reduce risk of heart problems in obese or overweight adults, July 2024.
44. S. Hristakeva, J. Liankonyte and L. Feler, *The No-Hunger Games: How GLP-1 Medication Adoption is Changing Consumer Food Purchases*, Cornell University, 2024.
45. C. Doering, Nestlé launches drink that suppresses hunger, promotes GLP-1 production, *Food Dive*, 2024, Published Dec. 18.
46. W. K. Viscusi, The global COVID-19 mortality cost report card: 2020, 2021, and 2022, *PLoS One*, 2023, **18**(5), e0284273.

Epilogue

I started this book with a quote from Heraclitus, who believed that change is the only constant in life. The same philosophy centres on the idea that everything is in constant flux and that change also becomes essential for growth and transformation.

When we look back at human change over the past three million years, our physical evolution, our societies, the way we live, and the food we eat—all of that philosophy rings true. However, in more recent times, perhaps the one element of change that has affected us all is the continually increasing pace of societal change.

We have observed this phenomenon since the origin of our species, from Hominins, some 3 million years ago, to the first use of controlled fire and cooking, approximately 1 million years ago. *Homo sapiens* evolved some 300 000 years ago, and the agricultural revolution only began some 10 000 years ago. 2000 years ago, the first wind and water mills emerged, and 200 years ago, the steam locomotive became the driving force behind the Industrial Revolution. One hundred years ago, in the early part of the last century, there were the automobile, the first planes, synthetic fertilisers, and the invention of antibiotics, including penicillin.

The latter part of the 20th century saw the first computers, the moon landing, and the first eradication of human disease,

Epilogue 227

smallpox. It was not until 1991 that the Internet emerged, with the release of the first web browser and the debut of the first website. 1998 saw the birth of Google, and smartphones became ubiquitous.

In addition to the profound technological changes that have occurred relatively recently or because of them, the world's global population has quadrupled in the last century from 2 billion to 8 billion people. Longer lifespans and increasing urbanisation have become a new trend in reshaping the world and will continue as populations age and cities grow.

Meanwhile, our amazing bodies, which evolved to solve the survival problems of our Stone Age ancestors when humans lived in small-scale, hunter-gatherer societies, are still adjusting to the modern age. Our brains continuously put us at a disadvantage by responding to stimuli that—in prehistoric times—would have prompted behaviour that was beneficial to our survival but now clashes with our biology and modern culture.

It has been almost six years since I started researching this book from a "Food and Us" perspective. Food preferences and the start of the industrial food era have always fascinated me, as has my involvement and career development in the food industry. However, as I got further into the research, more questions arose. Why do we have a propensity for sweeter foods? What factors have determined our preferences for various food choices? Was it physiological, psychological, social, or genetic factors? Of course, the more I tried to answer these questions, the more I realised that the story of "Food and Us" goes way back to Hominin days when we first learned to walk on two legs.

Likewise, our current food situation, the growth of the food production business, and our current food systems. Has it become all about producing a suitable return on investment for the manufacturer or providing an essential food requirement for a growing population?

When I think of some of the past food projects I have been involved with, the latter question is not so easy to answer. Was creating an animal feed plant from the waste products of sugar cane milling, molasses, and bagasse (the waste cane fibre after juice extraction) better for animals or the bottom line of the sugar mill? Feed-grade molasses is a supplement commonly used to enhance palatability and increase sugar levels in livestock, including beef cattle, lactating dairy cows, and dry dairy cows.

Developing a wet milling process for producing High-fructose corn syrup on behalf of a maize grower's cooperative as an alternative, cheaper sugar source for Coca-Cola.

Working with commercial bakery suppliers to produce novel bread products such as "Sugarloaf" (18 percent sugar content!) that caramelised when toasted. Or a new "Fresh" branded loaf that could retain its freshness characteristics for a week or longer.

All of the above were new food products designed to increase the manufacturer's market share and profitability as "value-added" products. Still, they had little or nothing to do with improving the health benefits of the end consumer.

This was the eighties and nineties, and it was part of my personal food industry history. Knowing what I know now about health and well-being attributes, it would probably be best forgotten!

Unfortunately, we still see the ramifications of similar food "developments" across our global food supply. The quest to produce highly palatable, cheaper consumer food products with all the enhanced characteristics of our evolved food preferences combines with the economic incentives of food companies that prioritise volume, turnover, and profit and capitalise on satisfying this demand.

Likewise, the "Our animal-sourced meat dilemma" section outlines the increasing demand for animal-sourced meat and products, as well as the global ramifications of producing them.

Although global environmental concerns first surfaced with the counterculture movement of the late '60s and '70s, it was only in 1988 that global warming and ozone layer depletion became prominent in international debate. The first Conference of the Parties (COP) series on climate change began in 1995. The eventual Paris Agreement on climate change (COP 21) adopted on December 12, 2015, only came into force on November 4, 2016.

Around the same time, in September 2015, the United Nations (UN) adopted the Sustainable Development Goals (SDGs) to lead the world towards a more sustainable and equitable future by 2030. Upon further examination of these 17 SDG goals, it becomes clear that they are all related to "Food and Us" in some way.

Epilogue

The UN 2023 progress report emphasises "Times of Crisis, Times of Change: Science for Accelerating Transformations to Sustainable Development." It states that, at this critical midpoint toward 2030, incremental and fragmented changes are insufficient to accomplish the Sustainable Development Goals (SDGs) within the remaining seven years. Clearly, all stakeholders and governments need to work more concertedly to create a better world for all.

On a brighter note, from an individual or feeding a family perspective, given the proliferation of research and subsequent reading material available, we now have a much better understanding of how our evolved bodies function and their relationship with the food we eat. The intricate interplay between our genetic and physical makeup, our enteric nervous system and how our brain functions in unison with our microbiota to regulate how our body digests food.

Leaving industry marketing efforts aside, we also know what foods and portion sizes are best for creating a healthy, balanced diet: whole fruits and vegetables, whole grains, protein from limited red meat, fish, beans, nuts, and eggs, dairy or dairy alternatives, and water.

From a future food perspective, as part of the Food Systems Institute at the University of Nottingham, ground-breaking work and research are being conducted to ensure access to palatable, healthy, and sustainable food for all, while protecting and regenerating the Earth's natural resources in the face of climate change. The same effort is being mirrored by similar institutions across the globe, working towards common goals to achieve more sustainable and healthy food systems at local, national, and global levels.

Another positive note is how our younger generation, "Gen Z," recognises the urgency of our current situation regarding environmental matters and global sustainability. As part of my current role at the University, I have travelled to several countries across the globe to interview students for postgraduate studies, primarily in the fields of Chemical, Environmental, and Food Engineering. What intrigues me and gives me great hope for a brighter future is that almost 90 percent of those interviewed, when asked why they want to pursue a future career in these fields, reply in summary, "I want to help save the world"!

If you found the insights and themes explored in this book to be engaging, I warmly invite you to connect with me on LinkedIn for ongoing discussions and updates about our dynamic food system. You can easily locate my profile by searching for "Seamus Higgins Food Engineering."

I would also genuinely value your feedback on the book, whether it be praise, critique, or constructive suggestions. Your thoughts are important to me, and I would encourage you to share them by leaving a comment on the book's dedicated website at www.foodandus.life.

I'm also excited to announce that I plan to write and publish some additional articles and blogs centred around my passion for food and our intricate food system. I would be delighted if you chose to register for these upcoming pieces, as I believe they will provide further valuable insights into this vital topic.

Last but not least, on a more personal note, the sleeve cover for "Food and Us" asks, "What are your favourite foods?" It then describes some of my favourite foods while growing up in the 1970s. It would be remiss of me, having dedicated the book to my children and grandchildren, who know me better, not to admit that while I have managed to ditch Angel Delight and instant potatoes, for better or for worse, I am still very partial to ice cream and Arctic rolls ☺.

Subject Index

ABCDs of grain trading 45, 176
Aboriginal Australians 65–66, 218–219
Abraham, Danny 131
addictions 74, 156, 165, 220–221
    morphine 4, 97
additives 114, 154–156
    bread-making 62
adenosine triphosphate (ATP) 147
adolescents
    obesity 186–187, 188
    ultra-processed foods 190, 203
Adulteration of Food and Drugs Act in 1872 113
advertising 98–106, 204, 205
    sugar industry 120–121
Africa, hominim evolution 12, 13, 14, 16, 19, 20
agriculture (farming) 29–55, 80–93
    First Agricultural Revolution 29–42
    industrialised 80–93
    prices 132
    Second Agricultural Revolution (Britain/UK) 69, 226
Agri-Food sector and systems 5, 138
    costs 6, 178
air, technology to produce food from 215, 216, 217
Air Protein Company (California) 216
Akkadian 36, 37
Alazani Valley 67
alcohol 98
America *see* Central America; Mesoamerica; New World; South America; United States of America
amylase 62
animals
    animal-sourced food (ASF) 17, 91, 178–184, 228
    breeding 82–83
    domestication 33, 35, 85, 214–215
    draft 33
    foraging 151

animals (*continued*)
    gelatin as byproduct from 103, 179–180
    homeostasis 25, 138
    humans and other animals, similarities and differences 150
    as meat source *see* meat
    milk *see* milk
    welfare 87–88
animism 36, 56
antibiotics/antimicrobials 88–90, 219
    factory farming 88
    resistance (incl. superbugs) 89–90
Anunnaki 36
Appert, Nicolas 69, 189
Archer Daniels Midland company 176
arterial atherosclerosis 118
atherosclerosis 118
Atlanta's prohibition laws (1885) 98
ATP (adenosine triphosphate) 147
Atwater, Wilbur 157
Australians, Aboriginal 65–66, 218–219
autonomic nervous system 25

babies and infants 187–190, *see also* breastfeeding; parents
    convenience food 189, 190
    feeding/diet 187–189
    newborn microbiome 147–148, 188
    parental eye movements and 22
    squeeze pouch foods 189–190
baby boomers 124, 207, 208
bacteria and oral environment and teeth 34, *see also* microbiome
bagasse 227
baguette 59–60
baker's yeast 38
barcodes 109
barley (barley grain) 35–36, 45, 47, 49
Baum, Rudy M. 126
Bayer 85
beef
    beef fat (for spreading) 114–115
    cooked 179
beer, monasteries 68
"Behind the Brands" (Oxfam campaign) 135–136
beriberi 50–51, 52
Berkshire Hathaway investment 176
Berthelot, Pierre 157
beverages *see* drinks
"Big Food" players 134, 135
biological factors with food preferences 179
BioMed Central (BMC) Public Health (Australia) 190
bipedalism 12–14
Birdseye 102
Bite Back's "Fuel Us, Don't Fool Us" campaign 136
black-tongue disease (dogs) 53
"bliss point" 1, 166–167

Subject Index 233

BMC Public Health (Australia) 190
bomb calorimeter 157
Borlaug, Normal 82–83
Boston Consulting Group's (BCG) growth share matrix 96
bovine spongiform encephalopathy (BSE) 116
Braconnot, Henri 77
brain 15–18, 23, 24, 25, 148–149, 166, 167, 168, 227
    gut–brain axis 148
    limbic system 24
    prefrontal cortex 24
    size 14, 15, 17, 178–179
brands 204, 214
    niche 135
Brazil 75, 160–161, 181, 182
bread 56–63, 228
    unleavened (flatbread) 56, 57, 58
    white 60–61, 61–62
breakfast cereals 99, 101, 203
breastfeeding 1, 22, 187
    breast milk 40, 166, 187, 188
breathing 26
breeding
    animals 82–83
    plant 82–83, 214–215
    selective 69, 82, 83
Britain *see* United Kingdom
British Association for the Advancement of Science (1898 meeting) 80–81
British Baking Industries Research Association (BBIRA) in Chorleywood 61–62, 63

British Medical Journal and ultra-processed food 165
British Nutritional Institute 159, 164, 169
Broca's area 15
broilers 86
Bronze Age 31, 37
brown rice 8, 50, 51, 52
BSE (bovine spongiform encephalopathy) 116
Buffet, Warren 176
Buhler, Adolf 49
Bunge 176
burgers and hamburgers 104
Burnett, James 10
Burnett, Leo 100
businesses and companies, *see also* food industry
    consolidation 133
    economic implications of future policies for 197
    management theories in food production 3, 94–97, 172 173
    purpose 3
butter *vs.* margarine 106, 114–115, 117, 160

Cadbury, John 76
calcium 151
    white flour 115–116
calories 156–160
    counting 13, 158, 161
cancer 117
    obesity and 191
Candler, Asa 98
canned foods (tin cans) 70, 154
    meat 102–103

capitalism 3, 48
  shareholder 96
  stakeholder 3, 96
carbohydrates (carbs) 34, 122, 126, *see also* fibre
  dopamine and 166
  highly processed/refined 123, 146
  starch-based 130
carbon emissions *see* climate change; greenhouse gas emissions
Cargill 176
Carlsson, Arvid 167
carnivores 17
Carson, Rachel, *Silent Spring* 125–126
Catholic (Roman Catholic) church 57, 67, 68, 72
cattle, early societies 36
Center for Science in the Public Interest (CSPI) 202
Central America, maize nixtamalisation 53
central nervous system 148, *see also* brain
cereal grains *see* grains
Ceres (Roman Goddess of Grain and Agriculture) 44, 63
CFCs (chlorofluorocarbons) 201–202
changes
  in food systems
    costing 196–202
    market changes 131–136
  Heraclitus's only constant is change 1, 7, 26, 198, 230

Cheskin, Louis 4, 104–105
Chicago Board of Trade 174
chickens 86
Childe, V. Gordon 31
children, *see also* adolescents; babies and infants
  healthy foods 189
  human evolution and 22
  Marshmallow test 24
  number of (in family) 22
  obesity 186–188
  schoolchildren 212–213
Chile, ultra-processed food 190
chimpanzees 13, 151, 186
China
  ancient times 30, 37, 66
  broiler industry 86
  meat 181
  rice milling 50
Chinese language and writing 8, 50
chlorofluorocarbons (CFCs) 201–202
chocolate and cocoa 2, 76–77, 166
cholesterol 118, 186, 219
  statins lowering levels of 123, 219, 221
Chorleywood (village), British Baking Industries Research Association in 61–62, 63
Chorleywood Bread Process (CBP) 62–63
Christianity 2, 57, 68, 100
chuggura (rainstick) 218
Cities and the Circular Economy for Food 138
Clement, Nicolas 156–157

climate change and global warming 181–182, 200, 217, 228, *see also* greenhouse gas emission
Coca-Cola 98, 176–177, 204
cocaine 97
cocoa and chocolate 2, 76–77
cognitive abilities 178–179, 186, 197
Cohen, Jack 107
cold plasma 218–219
colour 21, 101, 115
    colour-coded food packaging 158
    eye 7, 144
    flavour and 101
    McDonalds 4–5
Columbus and the Columbian Exchange 72–74
combine harvester 81
commodities 174–175
companies *see* businesses and companies
compound annual growth rate (CAGR) 4, 106, 135
ConAgra company 221
condiments 77–78
Conference of the Parties (COP) series on climate change 228
Confined Animal Feeding Operations 87
Confucius 8
consolidation (in food industry) 133
convenience foods 4, 97–106, 134, 154, 171, 207
    babies 189, 190
cooking 19

COP (Conference of the Parties) series on climate change 228
corn (maize) 30, 44, 52, 53, 182
    as agricultural commodity 174
Corn Flakes (Kellog's) 99
corn syrup 155
    high-fructose (HFCS) 106, 129, 228
coronary heart disease (CHD) 117–123
Cortés, Hernán 76
costs 196–202
    of Agri-Food sector and systems 6, 178
    of COVID-19 pandemic 222
    environmental 6, 138, 199
    of food 131, 199
    of food system changes 196–202
    health 191, 196, 199
counterculture movement 124–126, 129, 178, 202, 228
COVID-19 pandemic 90–91, 110, 139, 148, 222
    global cost 222
    grain trading during 175, 176
Craig, Jenny 131
Crick and Watson 142
Crookes, Sir William 80–81
crops 82–85
    domesticated 29, 30
    varieties/cultivars 82, 83, 210
    yield improvements 80, 81, 218
Crutzen, Paul 201

CSPI (Center for Science in the Public Interest) 202
culinary ingredients, processed 162
Cullen, Tracy 212
culture and food preferences 179
currency *see* money and currency

Da Vinci, Leonardo 45
Dagher, Alain 166
daily life in human evolution 22
dairy industry 87, 114
Danone 135
Darius (Emperor) 74
Dart, Raymond 12, 16
Darwin, Charles 10–12, 12–13, 30, 142–143
deforestation 130, 182
Deloitte 110
Demeter (Greek Goddess of Grain and Agriculture) 44, 56–57, 63
Dene, Oliver Stone 95
Denisovans 19–20
dentition (teeth), evolution 13, 17, 34
Descent of Man (Darwin's) 11, 30
developing (low- and middle-income) countries 133, 135, 138, 181, 187, 198
diabetes 5, 123, 219, 220
Diamond, Jared 85
diet
  evolution 149
  guidelines 121, 122, 164
    babies/infants 189
  high-fibre 61, 149
  low-fat diet 121, 123, 129–130
  Mediterranean 184, 190
  mismatch of ancient physiology and modern diet 146
  Paleo 15–16
  plant-based
    see plant-based diet
diet food 129–131
digestive system *see* gut
Digneau, Ken 103
Dimbleby, Henry 206, 211
diseases
  of civilisation 146
  GLP-1 agonists and 220
  infectious *see* infectious disease
  microbiome and 147
  sugar and fat and 117–123
DNA 142–145, 147, 150–151
  junk 144–145
  mitochondrial 147
dogs, black-tongue disease 53
domestication
  animals/livestock 33, 35, 85, 214–215
  plants/crops 29, 30, 214–215
Donkin, Bryan 70
dopamine 146, 165–166, 168, 203
draft animals 33
dried cereal products
  babies/infants/toddlers 190
  Post (Charles W.) in 1897 99

Subject Index 237

drinks/beverages 188
    soft 4, 122, 203
        sugar-sweetened 188, 208, 211
Drosophila (fruit flies) 38–39
Drucker, Peter 3, 95
drug treatment of obesity 220–221
due diligence 116
Dunnhumby 109, 110
Dyson, Dr Lisa 215, 216

E numbers 155
Earth, and the Gaia hypothesis 137–140
Earth Day 125
Eastern Orthodox church 68
Ebola 91
economics/economies
    economic growth 178, 197
    economic impact of future policies on farmers/consumers/businesses 197
    economic impact of obesity 191
    economies of scale 134
education 212, 213, 214
eggs 86
Egypt, ancient 37, 82
    bread 57
    wheat varieties and cultivation 82
Eijkman, Christiaan 51
einkorn 29, 43, 210
elephants 150
Ella's kitchen 189
Ellen MacArthur Foundation 6, 135, 138
emmer 29, 43, 210

ENCODE (ENCyclopedia Of DNA Elements) 145
Enders, Giulia 149–150
energy 154–156
    animal-sourced food/meat 18, 179
    fat as reserve of 146
    labels 156–160
English Reformation 68
enteric nervous system 8, 148, 229
environmental factors and concerns 5, 135, 137–140, 199
    counterculture of 1960s/1970s 125–126
    environmental costs 6, 138, 199, 219
    sustainable environment 125, 138, 200, 213
enzymes in bakery trade 62
Europe
    E numbers 155
    origins of farming 30
European Union (EU), reference intake table 158
evolution (human) 1, 10–28, 31, 142, 226
    dopamine and 167
    from hunter-gatherers 172
expensive tissue hypothesis 17–18, 178–179
eyes
    colour 7, 144
    human evolution and 21–22

Facebook 208
factory farming 86, 87–88, 139, 181

FADS1 and FADS2 enzymes 144
FAO (Food and Agricultural Organisation) 6, 116, 178
farming *see* agriculture
fast foods 5, 105, 105–106, 204
Fast Moving Consumer Goods (FMCG) 110, 180
fats and fatty acids 117–123, *see also* HFSS (high fat, salt and/or sugar products); oils
    beef fat (for spreading) 114–115
    dopamine and 166
    as energy reserve 146
    high-fat diet 123
    hydrogenation of vegetable oils into solid fats 106, 115, 130
    low-fat diet 121, 123, 129–130
    omega-3 and omega-6 fatty acids from plant-based diet 7, 144
    saturated 117–123
    short-chain (SCFAs) 149
FDA (US/USA – Food and Drug Administration) 115, 155, 156, 158, 219, 220
fermented foods
    ancient times 38–39
    microorganisms in production 38, 215
Fertile Crescent and Mesopotamia 29, 33, 172, 180
fertilisers 216–218
    chemical 80
    environmentally friendly 216–217
    synthetic 139, 217

feudalism 47
fibre 139
    children 208
    high-fibre diet 61, 149
fiction, shared 214
fight-or-flight response 25
Finland, Solar Foods 216
fire 19
fish 180, 181
    ancient humans 19
flatbread 56, 57, 58
flavours *see* taste
Flavr Savr tomato 84
flexitarian diet 184
flour
    white *see* white flour
    wholemeal 116
fluids/liquids 210, *see also* drinks
FMCG (Fast Moving Consumer Goods) 110, 180
folic acid fortification 117
food(s), *see also* diet; nutrition
    additives *see* additives
    advertising *see* advertising
    advice (20th/21st century) 113–117
    animal-sourced 17, 18, *see also* meat
    "bliss point" 1, 166–167
    brands *see* brands
    as commodity 174–175
    convenience *see* convenience foods
    costs 131, 199
    as currency, early societies 35–36
    healthy *see* health foods
    junk 202–209

Subject Index 239

new (in ancient times), developing 37–40
NOVA food classification system 62, 159, 160–165
origins 2
packaging *see* packaging
preferences/favourite 207, 208
   author's 230
   babies 190
   calorie-rich foods 206
   meat 179, 181
   regional 65–71
processed *see* processed food
production in 20th century 94–112
regulations *see* law and regulations
reshape consumer demand 221
safety *see* safety
storage and preservation *see* storage and preservation
supply 3, 171–178, 228
surplus, early societies 35, 38
waste 5, 138–139
whole 152, 163, 202
Food and Agricultural Organisation (FAO) 6, 116, 178
Food and Drug Administration (US) 115, 155, 156, 158, 219, 220
Food Beast 101

food industry 3, 6, 131–136
   business management 3, 94–97, 172–173
   changing market 131–136
   law *see* law and regulations
   profitability 4, 131–136
food literacy 212
food system(s) global 138, 171–195
   changes *see* change
   supporting a new generation 209–214
   sustainable 5, 125, 137–140, 183, 196–226
Food Systems Institute (University of Nottingham) 229
Food System Economics Commission (FSEC) 198–199, 199
foot traffic flow patterns 108
Ford, Henry 95
forest destruction (deforestation) 130, 182
fortification 115, 117
   of white flour 115–116, 117
France (and the French) bread 59–60
   grain milling (in 1369) 48
   Nutri-Score label 159
Freud's theories 3–4, 105
Friedman, Milton Friedman 3, 96, 172–173
Froot Loops 100–101
fruit flies 38–39
Fry, Joseph 76
FSEC (Food System Economics Commission) 198–199, 199

"Fuel Us, Don't Fool Us"
  (Bite Back's campaign) 136
fungal toxins (mycotoxins) 54
fungicides 80
Funk, Casimir 51
future (the)
  awareness/thinking of
    23–24
  sustainable systems in
    196–225

Gaia hypothesis 137–140
Gauls 59
gelatin 103, 179–180
gene(s) 143
  mutations *see* mutations
General Electric's (GE)
  strategic planning 96
Generation A (alpha) 209–214
Generation X 207, 208
Generation Z 208, 210, 213, 229
genetic engineering (including
  GM/genetic modification)
  83–84
genetics (incl. genetic variants)
  142–146, 187
  vegetarian diet and 7, 144
genome, human 143, 144–145,
  150
genotypes 145–146
Georgia, winemaking
  37–38, 67
Germany rye bread 58
Giphart, Ronald 22
Global Action Plan for the
  Prevention and Control of
  Non-Communicable
  Diseases 210
global food systems *see* food
  systems

global populations 2–3, 85,
  131, 184, 198, 199, 208, 227
global sustainability 197
global warming *see* climate
  change and global warming;
  greenhouse gas emissions
globalisation 3, 80, 86, 96, 116,
  171
GLP-1 agonists 220–221
glyphosate 84–85
Göbekli Tepe 31–32
Goldberger, Joseph 52–53
golden arches (McDonald's)
  104, 105
gorillas 15, 151, 185, 186
grains (cereals)
  ancient 210
  dried, babies/infants/
    toddlers 190
  early societies 35
  future generations 210
  milling *see* milling
  most commonly
    consumed 44
  storing and shipping 44
  trading 45, 174–176
granola 99, 124
grassland savannah 16
gratification, instant and
  delayed 24
great apes 185, 186
Great Britain *see* United
  Kingdom
Greeks (ancient) 2, 10, 36, 39,
  44, 137
greenhouse gas emissions 5, 138
  animal agriculture 181–182
  cultured meat and 215
  fertilisers and 217
Grill-Yang price index 132

growth, economic 178, 197
growth share matrix (BCG) 96
Guam 103
Guideline Daily Amounts 156
gut (digestive system) 147–150, see also enteric nervous system
    gut–brain axis 148
    microbiome see microbiome

H1N1 and swine flu pandemic 90
Haber–Bosch process 81
HACCP (Hazard Analysis and Critical Control Points) principles 116
Hadley, Mr Simon 50
ham, spiced (spam) 102–103
hamburgers and burgers 104
Harari, Yuval 13, 31, 214
Harvey, John 99
Hazard Analysis and Critical Control Points (HSCCP) principles 116
HDL (high-density lipoprotein) 118
health (physical and mental health)
    costs 191, 196, 199
    public, recommendations 210
    sustainability and 207
health foods/healthy foods (today's) 124
    children 189
    government interventions 211
heart disease (coronary – CHD) 117–123

Heinz company merger with Kraft 133, 176
Heinz ketchup 77–78
Heraclitus 1, 7, 26, 198, 230
herbicides 84
HERC2 gene 144
HFSS (high fat, salt and/or sugar products) 101, 118, 130, 206, 209, 211
high-density lipoprotein (HDL) 118
high fat, salt and/or sugar products (HSSF) 101, 118, 130, 206, 209, 211
high-fat diet 123
high-fibre diet 61, 149
high-fructose corn syrup (HFCS) 106, 129, 228
high hydrostatic pressure 219
highly-processed foods see ultra-processed foods
hippies and counterculture movement 124–126, 129, 178, 202, 228
Hippocrates 31, 219
historical factors, meat and food preferences 180
home delivery 109
homeostasis, humans/animals 25, 138
hominins, evolution to *Homo sapiens* from earlier forms of see evolution
Homo erectus 15, 17, 19, 185
Homo genus, nomenclature 184–191
Homo habilis 15, 16–17
Homo longi 20

Homo neanderthalensis
  (Neanderthals) 19, 20, 185,
  185
Homo rhodesiensis 19
Homo sapiens
  evolution from earlier
    hominins *see* evolution
  nomenclature 184–191
horizontal transfer of
  antimicrobial resistance
  89–90
hormones 25, 187
Hu, Frank 122
human(s)
  evolution (hominins to
    Homo sapiens)
    see evolution
  homeostasis 25, 138
  mismatch of ancient
    physiology and modern
    diet 146
  similarities and differ-
    ences between other
    animals and 150
Human Genome Project 143
hunting 18, 22
  and gathering/foraging
    13, 19–22, 30, 32, 33,
    34, 37, 40, 56, 172
      Aboriginal
        Australians
        (today) 65–66
      quality of life and 33
hydrogenation of vegetable oils
  into solid fats 106, 115, 130
hypothalamus 25, 167

immune system and gut
  microbiome 148, 149
impulse purchases 109, 206

India, bread 58
Industrial Revolution 3, 69, 80,
  126
  end 5
  Fourth (Industrial 4.0)
    134
industrialised agriculture
  (of 20th century) 80–93
industry *see* food industry
infants *see* babies and infants
infectious disease
  (and pathogens) 89, 91, 148,
  see also antibiotics/
    antimicrobials
  20th century 113
  early humans 33
  New World 72–73
inheritance 12
innovation, technological 214
instincts 22–26
insulin 123, 219, 220
International Organization for
  Standardisation (ISO) 116
International Rice Research
  Institute (IRRI in
  Philippines) 82–83
Internet *see* online
intersexual selection 23
intestine 17
iodine supplementation/forti-
  fication 115
iron 179
Iron Age 31, 37
ISO (International Organization for Standardisation)
  116

Jacobson, Michael 202
Jam 77
Jericho 32

Subject Index

Jewish religion 57
Jobs, Steve 126
junk DNA 144–145
junk food 202–209, *see also* ultra-processed foods

Kakheti 67
Kantar 110
Kellog, John Harvey 99–100
Kellog's (business) 99–101, 204, 205
Kennedy, Ted 120
ketchup 77–78
Keys, Ancel 118–119, 123
Kluyveromyces lactis (*K. lactis*) 38, 39
Kluyveromyces marxianus (*K. marxianus*) 38, 39
Korten, David 178
Kraft–Heinz merger 133, 176
Kroc, Ray 104, 105
kvevris 38

labels and labelling 155–156, 158–159
lactose
    intolerance (lactase lack/deficiency) 144
    tolerance (lactase persistence) 39, 40, 172
land use 183
Landers, Johnathon 124
language
    Broca's area and 15
    Chinese, and writing 8, 50
law and regulations 113–117
    medieval (Britain) 58
    packaging 158
LDL (low-density lipoprotein) 118

Leonardo Da Vinci 45
lifespans, longer 227
limbic system 24–26
Limits to Growth 178
Lindt, Rudolf 76
Linnaeus and and Linnean system 10, 184–185, 185
liquids/fluids 210, *see also* drinks
livestock 85–91, 181
    early domestication 33, 35, 85, 214–215
long Covid 90
Louis Dreyfus Company 176
Lovelock, James 137, 139
low- and middle-income (developing) countries 133, 135, 138, 181, 187, 198
low-density lipoprotein (LDL) 118
low-fat diet 121, 123, 129–130
"Lucy" (human fossil) 12, 13, 14, 15, 142, 171
Luther, Martin 68
lycopene 155

Macdonald, Prof. David 90–91
McDonald's 4–5, 103–106
McGovern, George (Senator) 120
machinery for farming 81
MacLaurin (Lord) 109
Macron, Emmanuel (French President) 59–60
mad cow disease (bovine spongiform encephalopathy; BSE) 116
maize *see* corn

malnutrition (incl. undernutrition), *see also* overnutrition
   elimination by 2050 200
   low- and middle-income countries 187
   US 120
Malthusian theory 3, 80–81
mammals (incl. humans), common ancestor 150
management (in food production) 3, 94–97, 172–173
margarine 106, 114–115
   butter *vs.* 106, 114–115, 117, 160
Margulis, Lynn 137
Mariani, Angelo 97–98
market changes 131–136
marketing and the future 207
marmalade 77
Mars 76–77, 136
Marshmallow test 24
Massachusetts Institute of Technology (MIT) 126, 178
Meadows, Dennis and Donella 178
meat (animal-sourced) 178–184, 228
   canned 102–103
   cultured/cultivated 215
   dilemma (for the future) 178–184, 215, 228
   early humans 17
   as energy source 18, 179
mechanical farm machinery 81
medieval times *see* Middle Ages
Mediterranean diet 184, 190
mental health *see* health

Mesoamerica 2, 53, 65
Mesopotamia and Fertile Crescent 29, 33, 172, 180
methane emissions (ruminants) 181
Mexico 30, 53, 90
   maize nixtamalisation 53–54
   tax on non-essential food items 211
   ultra-processed food 190
mice 150
microbes/microorganisms (use)
   fermentations 38, 215
   new technology 214
microbiome (gut) 147–150
   newborn babies 147–148, 188
Middle Ages (medieval times) 40, 57, 58, 65, 66, 67, 68
   preferences of food and taste 65, 66, 67
Middle East (Near East) 29, 30–35, 37, 73
milk
   animal 7, 144, 210
      ability to digest 39, 144
      fermented 38–39
      pasteurised 114, 188
   breast 40, 166, 187, 188
Millennials 207–208, 208, 210
millet 30, 43, 44, 66
milling (grain) 43–54, 182–183
   wet 228
minimally processed foot 161
Mischel, Walter 24
MIT 126, 178
mitochondria 147

Subject Index   245

molasses 227–228
Molina, Mario 201
monasteries 66–67
money and currency 171–178, *see also* costs
    early societies 35–36
monoculture 83–84, 117
Monsanto 84–85
Monteiro, Carlos 62, 160–161, 164
Montreal Protocol 201
Morgan, Lewis Henry 30
morphine addiction 4, 97
Moskowitz, Howard 166–167
motivation 25, 167, 206
Mounjaro 220
mouse (mice) 150
music in supermarkets 108
mutations 7, 143–144
    process of creating (mutagenesis/mutation breeding) 83
mycotoxins 54

naan bread 58
Napoleon Bonaparte 59, 69, 75, 106, 189
National Human Genome Research Institute (NHGRI) 145
National Institutes of Health and ultra-processed food 164–165
Natufians 29
natural selection (Darwin's) 11–12, 143
Neanderthals (*Homo neanderthalensis*) 19, 20, 185, 185

Near (Middle) East 29, 30–35, 37, 73
Neolithic times, agriculture 29–42
nervous system
    autonomic 25
    central 148
    enteric 8, 148, 229
Nestlé (company) and Henri Nestlé 76, 131, 133, 134, 136, 173–174, 209, 221
New World 72–74
niacin (nicotinamide; vitamin B3) 51, 52, 53, 115, 116, 179
niche brands 135
nicotinamide (niacin; vitamin B3) 51, 52, 53, 115, 116, 179
Nidetch, Jean 130–131
Nile valley 37, 44
Nipah 91
nitrogen 217–218
    as alternative to Haber–Bosch 217–218
    to ammonia (in Haber–Bosch process) 81, 217
    pollution 199
nitrous oxide (N_2O) 217
nixtamalisation 53–54
No-Hunger Games, the 221
Norman Conquest 47
    after (post-Norman times) 38
Nottingham University Food Systems Institute 229
NOVA (food classification) 62, 159, 160–165
Nutri-Score label 159

nutrition 154–170,
  see also diet; food
    20th century deficiencies 115
    early agriculture and 33–34
    school health and 213
Nutrition Action 202
Nutrition Facts 156
nutrition labelling 156, 158

obesity (and overweight) 5, 123, 151, 184–191, 199, 206
    dopamine and 167–168
    drug treatment 220–221
Office International des Epizooties (OIE) 88
oils, see also fats
    as processed culinary ingredients 163
    vegetable, hydrogenation into solid fats 106, 115, 130
Oldowan 16
omega-3 and omega-6 fatty acids from plant-based diet 7, 144
online (Internet)
    marketing/advertising 101, 205
    shopping 101, 109, 110
orangutans 10
overeating/overnutrition 138, 151
overweight see obesity (and overweight)
Oxfam's "Behind the Brands" campaign 135–136
oxygen 26

Ozempic 219, 220
ozone layer depletion 201–202, 228

packaging 155, 158
    baby food in pouches 189
    labelling 155–156, 158, 159
Paleo-diet 15–16
parents and their infants
    eye movements and 22
    recent research 209
Paris Agreement on climate change 228
pasteurised milk 114, 188
pathogens see infectious disease
pectin 77
pellagra 52–53
Pemberton, John 97–98
penicillin 219
    resistance 89, 90
PepsiCo 136
personal intelligent technology 213–214
pesticides 80, 125, 139, 210
Peter, Daniel 76
phenotypes 145–146, 168
Philippines, International Rice Research Institute (IRRI) 82–83
physical health see health
physical makeup and changes 7, 142–153
    agriculture and 33–34
pig see pork; swine flu pandemic
Piggly Wiggly store 107
Pilnick, Garry 204–205
Plan, The 206, 211

Subject Index

Planet Tracker 132
plant-based diet 183–184,
   see also vegetarian diet
      omega-3 and omega-6
         fatty acids from 7, 144
plant breeding 82–83, 214–215
plasma technology 217–219
pleasure and dopamine 166, 167
Pliny the Elder 39, 45, 59
Plum Organics 189
Polmon, Paul 139
Pontzer, Herman 186
populations, global 2–3, 85, 131, 184, 198, 199, 208, 227
pork (pig meat) 87, 103, 181, *see also* swine flu pandemic
Post, Charles W. 99, 100
Post, Mark 215
poultry 86
poverty
   richness and,
      elimination 59
   structural 199
power for milling grain 49
precision fermentation 215
prefrontal cortex 24
preservation *see* storage and preservation
preserves 77–78
price 97–106
   agricultural 132
primates (non-human) 18, 21, 185, 186
   ancestral 13
   great apes 185, 186
probiotics 39, 40
processed culinary ingredients 162

processed foods (refined foods) 161–165
   babies and infants 188, 189, 190
   definition 159
   highly *see* ultra-processed foods
   minimally 161
   numerous refining steps 204
   preceding 20th century 101–102
profit and profitability 4, 131–136, 171–176
promotions on unhealthier foods 206
proteins 130, 151
   leverage 151
   produced from air 216
psychology
   evolutionary 23–24
   food preferences and 179
public health
   recommendations 210
pulsed electrical fields 219
pulsed UV 219

quick-service restaurant (QSR) and fast foods 5, 105, 105–106, 204

rainforest clearance 139
rainstick 218
Ralston, Johanna 191
Raubenheimer, David 151, 152
Rauber, Fernanda 155
Real Bread Campaign 63
reference intake table (EU) 158–159

refined foods *see* processed foods
regional food preferences 65–71
regulations *see* law and regulations
religion 1, 10, 29–33, 36, 180, *see also specific religions and monasteries*
    bread 56
Resale Prices Act (1964) 107
research and development departments 134
retort system (retorting) 70, 188, 189, 219
reward system 23, 25, 146, 166, 167, 203
rice 44, 51–52
    brown 8, 50, 51, 52
    Kellogg's toasted rice cereal 99
    milling 50, 52
    parboiled/converted 51–52
    white 51, 52
richness and poverty, elimination 59
Robinson, Frank 98
Rockström, Johan 196
roles within society 36–37
Roman Catholic church 57, 67, 68, 72
Roman times 36, 37, 44, 47, 49, 57, 58, 59
Roundup 84–85
Rowland, Frank 201
ruminants, methane emissions 181
rural to urban areas, move from (urbanisation) 3, 96, 138, 178, 196, 198, 228

Ruskin, John 92
rye 44
    bread made from 58

Saccharomyces cerevisiae (*S. cerevisiae*) 38–39
safety (food) 113, 114, 116–117, 154
    genetically modified products 83–84
salmonella 34, 116
salt 40, 151, 155, 211, *see also* HFSS (high fat, salt and/or sugar products)
    in bread-making 62
    as food currency 36
    iodine fortification 115
    in processed food 162
Sargon (of Akkad) 37
SARS (severe acute respiratory syndrome) 91
saturated fat 117–123
Saunders, Clarence 107
savannah grassland 16
Scappi, Bartolomeo 74
scavenging 17
Schmidt, Klaus 31–32
schoolchildren 212–213
Scientific Advisory Committee on Nutrition (UK) 159
Second World War (WWII) 61, 103, 103–104, 114, 115, 116
self-service store, World's first 107
semaglutide 219–220, 221
severe acute respiratory syndrome (SARS) 91
sexual selection 11, 22–23

sexuality
    dopamine and 167
    Freud and 4, 105
ShareAction 173, 174
shared fiction 214
shareholders 172–173, 173, 176
    shareholder capitalism 96
    shareholder value 172–173, 177, 177–178
Sheehan, Evan 110
shekel 35
short-chain fatty acids (SCFAs) 149
Silcock, Chris 101
Silent Spring (Rachel Carson) 125–126
Simon, Henry 49
Simpson, Stephen 151
single-cell organism or microbe 216
SLC24A5 gene 144
Slimfast 131
Smith, Adam 106
Smith, Richard "Stoncy" 61
smoking 118
social characteristics of livestock 85
social impact of future policies 197–198
social media 204, 208, 214
social relationships 21
society
    Medieval 66
    modern, stress 25
    neolithic 34–37
    roles within 36–37
sodium (blood) 40
soft drinks 4, 122, 203
Solar Foods (Finland) 216

Solein 216
Sonnenborn, Harry 104
sourdough bread 59
South America 50
soy/soy bean 182
spam 102–103
spiced ham (spam) 102–103
squeeze pouch products 189–190
stakeholder capitalism 3, 96
starch-based carbohydrates 130
statins 123, 219, 221
Steele, Cecile 86
Steele, Tanya 186
stomach 17
Stone Age 15, 227, *see also* Neolithic times
stone tools 16
storage and preservation 69–70
    fermented food 38
    grains (by Romans) 44
    Napoleonic era 69, 189
    new technology 218
strategic planning (GE) 96
stress 25
stroke 40, 117
Stroud, David 121
structural poverty 199
sugar 74–76, 117–123, 155, *see also* HFSS (high fat, salt and/or sugar products)
    drinks/beverages sweetened with 188, 208, 211
    "no added" 188, 189, 210
sugar cane 74, 75, 227
Sumerians 35–36, 37, 56
supermarkets (UK) 107–110

sustainability 5, 125, 137–140, 183, 196–226
    environmental 125, 138, 200, 213
    global 197
Swanson (US company) 102
sweeteners
    corn syrup *see* corn syrup
    synthetic 84
sweetness (taste for) 74–77
swine flu pandemic 90, 91
Switzerland
    chocolate 76, 77
    salt fortification with iodine 115
Symbiobe 217

Taggart, Alexander 60
tastes (and flavours) 97–106
    animal products 179
    colour and 101
    counterculture and 124
    of Froot Loops 101
    junk food 203
    new (in ancient times), developing 37–40
    regional preferences 65
    for sweetness 74–77
"Taung Child" 12
taxation 211
    salt 40
Taylor, Frederick, and Greater Taylorism 95, 97
technology 109, 134, 154, 209, 214–222, 226–227
    innovation 214
    personal intelligent 213–214
    sustainable 214–222
teeth, evolution 13, 17, 34

temperature (body) regulation 26
Tesco 107, 109–110
Tetrapak. 135
thermodynamics, first law of 157, 164
thermoregulation (body temperature regulation) 26
thiamine *see* vitamin B1
TikTok 205, 208
tin cans *see* canned foods
tirzapeptide (Mounjaro) 220
tofu 124, 183
tomatoes 73
    genetically modified (Flavr Savr) 84
Tony the Tiger 100
Toucan Sam 100–101
trading in grains, ABCDs of 45, 176
traffic light labelling system 158
transparency 134–135
transportation (incl. shipping)
    grains (by Romans) 44
    Middle Ages 66
Triticum aestivum see wheat
Turkey, origins of farming 31
TV dinners 102, 124

UK *see* United Kingdom
Ukraine (war in) 175, 176
ultra-processed (highly-processed) foods 5, 149, 151–152, 155, 159–160, 163–165, 164, 190–191, 202, 221
    carbohydrates 123, 146
    consumer spending on, reduction 221

Subject Index

definition 159
examples 163
in NOVA classification 62, 159, 163–165
ultrasonics 219
ultraviolet (UV), pulsed 219
umami 151, 179, 216
UN *see* United Nations
undernutrition
  see malnutrition
Unilever 114, 131, 136, 139
United Kingdom (Britain; GB; UK) 106–110
  bread-making and yeast production 60–62
  English Reformation 68
  government 101, 191
    1992–2020 policies 206
  junk food 203
  Kellogg's 101
  law and regulations 113, 114–115, 116, 117
    first 58
  lifestyle 118
  nutrition labelling 156, 158, 159
  retailing 106–110
  Second Agricultural (British) Revolution 69, 226
  sugar/fat (unhealthy/processed) consumption in 118, 119, 121, 159
    taxation 211
  supermarkets 107, 109, 110
  ultra-processed foods 164

United Nations (UN) 82, 175
  antimicrobial resistance and 90
  Convention on the Rights of the Child 209
  Educational, Scientific and Cultural Organization (UNESCO – 2022 report) 213
  Environment Programme 199
  Sustainable Development Goals adopted by 228–229
  "Times of Crisis, Times of Change: Science for Accelerating Transformations to Sustainable Development" 229
United States of America (US/USA)
  corn production 192
  dietary guidelines 121
  Food and Drug Administration (FDA) 115, 155, 156, 158, 219, 220
  Great Depression 82, 100, 106, 115
  pellagra in southern areas 52–53
  spam and its history 102–103
  yeast production 60
unprocessed food 161, 164–165
urbanisation 3, 96, 138, 178, 196, 198, 228
US *see* United States of America
UV, pulsed 219

vegans 183
vegetable oils
    hydrogenation into solid fats 106, 115, 130
    land for crops 130
vegetarian diet and vegetarianism 183, *see also* plant-based diet
    genetic variants 7, 144
Vin Marian 97–98
"Vital Pursuit" 221
vitamin(s)
    origin of the term 51
    white flour fortification 115–116
vitamin B1 (thiamine) 52, 116
    deficiency 51
vitamin B3 (niacin nicotinamide) 51, 52, 53, 115, 116, 179
vitamin B9 (folic acid) fortification 117
Vitruvius and Vitruvian Man 45, 46

walking and bipedalism 12–14
waste
    food 5, 138–139
    sugar cane milling 227
water 26, 210
watermills 49
Watson and Crick 142
Wegovy 220
weight gain *see* obesity
weight loss
    diets 130–131, 158
    drugs 220–221

Weightwatchers 130–131
wet milling 228
wheat (*Triticum aestivum*) 45
    demonising consumption 184
    milling 49, 182–183
white bread 60–61, 61–62
white flour 45–46, 49–50, 58
    fortification 115–116, 117
white rice 51, 52
WHO (World Health Organization) 91, 116, 187, 187
whole foods 152, 163, 202
whole grain cereals (future children) 210
wholemeal flour 116
windmills 49
winemaking in Georgia 37–38, 67
World Health Organization (WHO) 91, 116, 187, 187
World Obesity Federation 5, 191
World Organization for Animal Health (WOAH) 88
World Trade Organization (WTO) 88
World War II (Second World War; WWII) 61, 103, 103–104, 114, 115, 116
Worldwide Fund for Nature (WWF) 125, 186
Wozniak, Steve 126

yeast, baker's 38
Yudkin, John 119–120